The Ultraview Effect

The Ultraview Effect

WHAT WE CAN LEARN FROM ASTRONAUTS ABOUT AWE, HUMILITY, AND EXPLORING THE UNKNOWN

Deana L. Weibel

UNIVERSITY OF CALIFORNIA PRESS

University of California Press
Oakland, California

Cataloging-in-Publication data is on file at the Library of Congress.

ISBN 978-0-520-40952-1 (cloth : alk. paper)
ISBN 978-0-520-40953-8 (ebook)

Manufactured in the United States of America

GPSR Authorized Representative: Easy Access System Europe, Mustamäe tee 50, 10621 Tallinn, Estonia, gpsr.requests@easproject.com

35 34 33 32 31 30 29 28 27 26
10 9 8 7 6 5 4 3 2 1

To my mother, who believed in honesty.

Contents

Illustrations and Table

ILLUSTRATIONS

TABLE

Introduction

With the corners of his eyes crinkling with joy at the happy memory, retired astronaut "Ben" told me about a moment of awe. A colleague who had already flown in space had given him some advice: During his space walk, while clinging to the outside of the International Space Station (ISS), Ben should stop for a moment and "just appreciate where you are." When that moment came for Ben, it was magnificent:

> I'm 220 miles above the Earth, nothing below me except the Earth. And I'm hanging on by a thumb and a forefinger and looking over my shoulder thinking, looking across—not *at* the Earth, but *past* the Earth—looking over the limb of the Earth,[1] out in the vastest universe and thinking, geez, there's nothing between me and *something* out there, you know. . . . I call it a goosebump moment. It was a really goosebump moment that I'm in this very, very unique position where there's *nothing*. The first time of my life there was *nothing*—looking not up but *across*—between me and the visible universe and just being amazed I had the opportunity to be in that spot . . . pointing, looking past the Earth. It made me more curious . . . what else is out there?

Ben experienced awe in the moment, followed by an intense feeling of amazement at his good fortune, and he was soon overwhelmed by curiosity.

Astronauts and others who explore outer space often experience a cycle I call the ultraview effect—moving from awe to humility to curiosity—especially when they leave Earth. This cycle helps them maintain perspective, stay open to new information, and adapt to changing circumstances. The ultraview effect uses amazement to allow one to reset perspective, discard false confidence, and acknowledge limitations, and encourages curiosity and cooperation, paving the way to new discoveries.

Being receptive to information that contradicts our existing beliefs is powerful and a crucial requirement for scientific exploration. A false sense of certainty, especially unexamined, closes minds. This unearned confidence can often be dismantled through experiences of awe. While it's possible to live without thinking much about what lies beyond Earth's atmosphere, those who regularly explore and study space must confront the vast reality of the universe. People I've interviewed who deeply consider what they don't know often describe feelings of awe, humility, and a desire to explore the universe's mysteries.

In her book *Blast Off! Train Like an Astronaut for Success on Earth*, Mindy Howard writes, "One of the unexpected effects of space travel is that many astronauts report emotional and spiritual transformation."[2] Throughout this book, which examines the different experiences astronauts have in space, Howard's words will prove to be both accurate and important. We'll learn, too, why these responses shouldn't be "unexpected."

At the time of this writing, America's National Aeronautics and Space Administration (NASA) is preparing to send humans to the

Moon for the first time in more than fifty years with the Artemis program. The last time people traveled to the Moon was during 1972's Apollo 17 mission, a mere eleven years after human spaceflight began with Yuri Gagarin's pioneering journey in April 1961. Since then, we have witnessed the rise and fall of Skylab, Mir, and the decades-long Space Shuttle program. We are now in a new era in which innovative vehicles like the SpaceX Crew Dragon transport astronauts to the ISS, while NASA's Orion spacecraft is preparing to take the Artemis II crew on a mission around the Moon.

When the Artemis II astronauts (and those who will follow) reach the Moon, they will have the rare opportunity to view the planet and the universe from a vantage point that is literally beyond what humans can experience on Earth. I expect that they will feel a kind of awe available only to those who, like some of the astronauts I've interviewed, confront the reality of how little humans understand the universe. This book offers an outer space–focused exploration of the importance of awe—how it fosters a positive and healthy sense of humility, and how humility, in turn, creates a foundation for learning and discovery. The types of awe humans experience in space differ from those experienced on Earth, but they provide valuable lessons for all of us.

People who study space, whether using telescopes or probes, or by sending crews into orbit, often function as a widespread but unified social group, sometimes known as the "space community." As a cultural anthropologist, I study human groups and their cultures in order to understand them. I have been researching the space community for more than twenty years, and what I have learned, especially from astronauts, has led me to write this book. Today, more and more people are experiencing things that until now have been limited to members of the space community. New technologies, like virtual reality and improved Hubble and Webb telescope photography, can give those of us on the ground a taste of what it is like to be up in space. Space tourism, too, is a burgeoning industry. Growing numbers of privileged private citizens are booking passage to experience otherworldly sensations like floating in microgravity or seeing the enormity of our beautiful blue-green planet with their own eyes.

While the technology and opportunities for people outside the space community have changed, space has long been a preoccupation of humankind. Many of the world's cultural and religious traditions see the sky through a religious lens, as the "heavens" or the realm of gods or ancestors. People journey to sacred places on Earth as well. I have often asked myself: Can space travel be understood as a kind of pilgrimage? Since the 1990s, I have studied pilgrimage and sacred places, mostly in Europe and North America, focusing on the wonder and humility often inspired by religion or religious experiences. It is not uncommon for a pilgrim at a sacred place like Mecca, Jerusalem, or the Ganges River to be awed by the encounter, overcome by a site's spiritual significance and humbled by a sense of connection with something that is otherworldly and beyond everyday human experience. I have noticed the same patterns among astronauts and other space professionals who contemplate the cosmos on a daily basis.

The way spirituality and human understandings of the cosmos intersect and influence each other fascinates me. Both space exploration and religion attempt to answer the big questions of where humans are and how we got here. The profound, almost spiritual, impact that space has on certain astronauts and other members of the space community is often overlooked by NASA and other space agencies. I want to share what I've discovered about the more humanistic and emotional parts of space exploration and what those of us on Earth can learn from these stories and experiences.

This book, however, isn't just about space and the people who explore it. It's about how humans approach the unknown in all aspects of life. Research suggests that leaders who project false certainty risk suppressing growth and innovation.[3] Politicians, CEOs, and sometimes parents make decisions based on limited knowledge without ever admitting they don't know everything, believing that it's better to project certainty than appear vulnerable. This viewpoint can lead to disaster, like impulsive dismissals of crucial personnel from nuclear plants and control towers, which jeopardized human safety, or the loss of human life that occurred when decision-makers in charge of the Challenger mission ignored the warnings of experts, leading to the deadly Space Shuttle explosion in 1986.[4]

But there is another way. Those who lead with openness and humility are better able to make constructive changes. Similarly, scientific breakthroughs require scientists to recognize gaps in knowledge and ask better questions. When people recognize their own limited understandings and are inspired to find out more, they can absorb new information more easily. Starting from a position of unexamined certainty leads to bad choices because it assumes that trusting what seems obvious is the only thing necessary for navigating a simple world. There is nothing simple about the world, its people, or our universe. True exploration—whether in space or in life—requires a recognition of complexity and a commitment to continued questioning.

I've learned from astronomers, engineers, flight surgeons, astronauts, and others in the space community that admitting one's own limited understanding in the face of complicated realities can lead to fulfillment and meaningful purpose, but also to a more positive mindset. The work required to keep humans alive and healthy during space travel, for instance, is an illuminating example because of the difficulty of that task. Medical issues result from radiation, and living in microgravity can damage bones, eyes, brains, and more. Despite this, "Nick," a circulatory system specialist who studies what microgravity does to the heart and blood vessels, told me, "I would say that most folks in space medicine say this is hard [but] somehow we can figure it out rather than giving up on it." These experts are intelligent people who stay open to possibilities and are eager for new data. Their optimistic view that solutions are possible and that openness and teamwork will allow humans to flourish in space in the future inspires the astronauts they work with, and I hope it will inspire you as well.

This book, then, is about how profound experiences like those of astronauts in space—experiences that connect awe and humility—can also occur on Earth, where they are capable of transforming our own perspectives, guiding us to a more honest and authentic approach to challenges and opportunities. Space exploration here is meant to serve as a sort of case study, highlighting creative and practical ways to embrace awe, cultivate humility, and nurture openness in ways that can translate to a variety of contexts on Earth.

My Own Journey to Space

I have always been drawn to stories about outer space, even those that were science fiction. Growing up in Southern California, I recall waiting in line for Disneyland's new Space Mountain roller coaster as a seven-year-old. Nervous but determined, I stayed despite numerous emergency exits, eager for a "high-speed, turbulent ride through space."[5] This experience gave me an early lesson about the thrill of exploration and the awe of the unknown. It was 1977, the year *Star Wars* premiered, and I stayed up late to catch a 10 p.m. showing with my dad. Sunday afternoons were spent watching *Star Trek* reruns. I was entranced by the idea of soaring through the stars and visiting different planets.

My enthusiasm for space and science fiction was soon joined by an interest in popular archaeology and anthropology, sparked by seeing King Tutankhamun's treasures in 1978 and *Raiders of the Lost Ark* in 1981. The ancient gods, goddesses, and supernatural elements filled me with awe and deep curiosity about our world's mysteries. The Jungle Cruise at Disneyland, with its imaginary Asian, African, and South American rivers, fueled my imagination, kindling my interest in different places, languages, religions, and cultures.

Of course, I didn't yet understand how amusement parks and entertainment aimed at children sensationalize and oversimplify space, science, and other cultures. As I realized that adventure movies and Disney rides were fantasy, I became humbled by how little I actually knew and decided to learn about the reality. Gradually, I became aware of the vastness of human experience and the importance of understanding it. I read everything I could—from *Omni* and *National Geographic* magazines to encyclopedias and almanacs. I set my alarm to watch Space Shuttle launches and landings, and grew increasingly intrigued by the diverse religious practices among my friends, visiting houses of worship and asking endless questions.

I studied linguistics as an undergraduate and pursued anthropology in graduate school at UC San Diego, where faculty specializing in psychological anthropology and religion shaped my perspective on how humans respond to profound experiences. In my twenties, I conducted field research in Rocamadour, France, a town with an ancient Catholic shrine built into the side of a cliff, inducing vertigo in pilgrims. This heightened my curiosity about how uncommon physical sensations impact spiritual feelings. Throughout, I continued to read science fiction and shared my interest in space exploration with friends, furthering my desire to investigate the unknown.

Two years after completing my doctorate, in 2003, I joined Grand Valley State University in Michigan, where I soon met Glen Swanson, whom I later married. Glen had recently returned from working as the chief historian at NASA's Lyndon B. Johnson Space Center and was founder of *Quest: The History of Spaceflight Quarterly*. He became my guide to the space community and remains my favorite "local expert." Through his connections, I gained a deeper understanding of NASA's awe-inspiring achievements and the humility outer space inspires in those dedicated to uncovering the universe's mysteries.

As my fascination with space exploration's cultural dimensions grew, I became curious about the historical and institutional forces shaping today's space community. The more I learned, the more intriguing parallels I found between astronauts' profound, otherworldly responses and spiritual experiences I had observed among pilgrims on Earth. These insights inspired me to develop the concept of the ultraview effect and motivated me to write this book. My intellectual interests and personal experiences significantly shaped how I understand these themes. While my cultural anthropology training strongly influences my perspective, I deeply value interdisciplinarity, drawing insights from psychology, history, biology, physics, and other fields. This interdisciplinary approach has helped me define and explain why awe, humility, and the human drive to explore matter profoundly—not just within the space community, but for all of us.

Who Gets to Be an Astronaut and Who Is This Book About?

Congress established the National Advisory Committee for Aeronautics (NACA) in 1915 to supervise aeronautical research. Shortly after the 1957 launch of Sputnik by the Soviet Union, Congress formed NASA, which absorbed NACA.[6] The first cohort of astronauts, the Mercury Seven, was composed of seven white men. The selection was announced on April 9, 1959, during a time when efforts were being made to solidify separate social roles for men and women in the United States. During World War II, while many men served overseas, women, often out of economic necessity, filled industrial and other jobs traditionally held by men. After the war, reconversion and veterans' priority in hiring resulted in the displacement of many women, and public messaging urged a return to homemaking.[7] That climate also shaped NASA's choices: the astronaut corps was drawn from the (then all-male) military test pilot pipeline.

Not everyone involved in space exploration saw it as a purely male occupation. Between 1960 and 1961, in a private study unaffiliated with NASA, Dr. William Randolph Lovelace II assessed women for their fitness for space, with many passing the same rigorous tests given to the Mercury Seven. The group of women who had proven themselves most space-worthy, known unofficially as the Mercury 13, were ready to go into space. In July 1962, however, a congressional hearing before the House Special Subcommittee on the Selection of Astronauts was held to determine whether women should be included in the U.S. space program. Members of the Mercury 13, including Jerrie Cobb and Janey Hart, argued for inclusion, amid rumors that the Soviet Union would be sending a woman into space. One criterion at the time, however, was test pilot experience in military jets, a qualification unavailable to women. Astronaut John Glenn, who spoke against women's inclusion in the program, said this requirement was crucial. The outcome upheld the status quo, and women were excluded.[8]

At about the same time that Dr. Lovelace was assessing women's fitness for space, a test pilot with military jet experience was being considered as America's first Black astronaut. Edward J. Dwight Jr., a U.S. Air Force aviator and aerospace engineer, was supported by President John F. Kennedy, who believed including a Black astronaut would send a powerful message of desegregation. Although Dwight was admitted to NASA's Aerospace Research Pilot School, he encountered significant institutional resistance. Some military personnel were resentful, as they perceived a Black candidate was being imposed upon them. Ultimately, Dwight was not selected for NASA's third astronaut group in 1963. Accounts vary regarding the role racial bias played in the decision. After Kennedy's assassination, Dwight lost his principal backer. Lyndon B. Johnson, who followed Kennedy as president, was less invested in promoting a Black astronaut than his predecessor, and momentum shifted. Eventually, in 1966, Dwight left the Air Force.[9]

Influenced by the civil rights movement, criticism of its lack of diversity, and the Equal Employment Opportunity Act of 1972, NASA later changed its requirements to include nonpilots. The new Space Shuttle could accommodate larger groups and required only two pilots per mission. NASA introduced a new category of "mission specialist" astronauts, which allowed doctors, engineers, and scientists without pilot experience to apply. In the 1978 astronaut class, women and people of color were included for the first time.[10]

Today, roughly 80 percent of American astronauts are white men, and this percentage may increase as NASA's diversity, equity, and inclusion (DEI) efforts are gutted under the Trump administration. Still, while most American astronauts are white men, the full cohort includes people of varied backgrounds. To more fully understand the space community and its culture, I interviewed a diverse sample of astronauts, engineers, technicians, flight surgeons, astronomers, and others whose work connects to space exploration. Among the astronauts interviewed for this book, more than 35 percent come from ethnically diverse backgrounds, including individuals from underrepresented racial and cultural groups in spaceflight. Moreover, astronauts from the Apollo, Skylab, Space Shuttle, and ISS eras are included among my participants, as are astronauts with scientific backgrounds and test pilots and military aviators.

NASA and other space agencies have worked toward greater diversity and inclusion, but recent political shifts have sparked concerns about whether these values will continue to be prioritized. Policies designed to increase representation face new scrutiny. Understanding this changing context helps us see why intentional inclusivity remains so important—not just because it's right, but because diverse perspectives make exploration and teamwork more creative, adaptable, and effective.

Anthropologists follow ethical guidelines that emphasize confidentiality and protection of interviewees' identities, especially when their experiences might make them targets for criticism or worse. Before interviews, we obtain consent from participants, promising confidentiality to the best of our ability. Although my research includes space professionals from diverse backgrounds, revealing specifics could jeopardize their privacy. Because women have become increasingly visible in space exploration, I felt comfortable mentioning gender. However, disclosing a subject's race, ethnicity, minority religion, or sexuality could easily identify an individual, so I've intentionally left those details out.

From Wonder to Humbleness to Curiosity

This book focuses on three main topics: awe, humility, and exploring the unknown. As we consider awe, we'll look at how different people around the world understand the concept and how it has been defined. We'll see how the awe inspired by experiences such as human-powered flight led to opportunities like space travel. Staring into a starry sky is one way people stir up wonder—astronauts experience something similar in space, although at a different level. The vistas from space travel are staggering, inspiring deep humility in those who experience them.

Our journey will continue with humility—how space exploration inspires it in the space community, and how various religions present humility as an important virtue. We'll examine how a lack of humility stifles creativity, while a humble attitude encourages curiosity and exploration. Humility may be the hardest step in the awe-humility-curiosity cycle—it's not as thrilling as awe or curiosity—but it's crucial because it allows us to recognize how little humans know and how much more there is to learn. Humility is essential to developing true curiosity.

Finally, we'll see that awe can transform human perspectives and make us more open to new information and experiences. By the end of this journey, we will understand that the lessons astronauts learn—connected to awe, humility, and curiosity—are valuable not just in space exploration but also in dealing with life's uncertainties on Earth. These insights can inspire a renewed appreciation of exploration, both around us and within ourselves.

The astronomer studying the Big Bang, the engineer flying probes to distant worlds, the doctor managing the physiological impacts of microgravity, and the astronaut staring at the stars all think deeply about outer space, but this doesn't mean they're detached from reality. Instead, they are aware of a reality many of us ignore: humans live on a tiny planet orbiting an average star in a medium-sized galaxy in a universe incredibly empty yet remarkably full of planets, stars, and galaxies. And, as hard as it may be to admit, we don't actually know much about it. Being aware of the universe's vastness is overwhelming, in a good way. As you read, let the awe and humility experienced by astronauts inspire your own exploration.

It's important for humans to experience awe—not just for its own sake, but because wonder enhances our outlooks, relationships, and clarity about reality. Awe's ability to inspire humility helps us avoid arrogance, breaking barriers of false certainty that block genuine exploration. I hope readers feel inspired by the astronauts' stories, like Ben's goosebump moment, to seek similar experiences here on Earth. Pursue what you don't yet understand, marvel at the unknown, appreciate how much more there is to learn, and embrace your capacity for discovery.

The Ultraview Effect's exploration of awe, humility, and curiosity will guide readers through significant insights drawn from the lives of astronauts and the space community. We start by examining awe itself, exploring how it shifts perspectives and opens us to new understandings. From there we move into humility, an invaluable trait for researchers pushing human knowledge limits. Finally, we consider how awe and humility inspire curiosity, motivating us to seek the unfamiliar. While the space community is central to this book, I conclude by sharing practical ideas to cultivate the ultraview effect right here on Earth.

I *Awe*

1 *The Essence of Awe*

Defining the Overwhelming

In 1952, British novelist Llewelyn Powys wrote, "No sight is more provocative of awe than is the night sky scattered thick with stars."[1] Back then it was easier to see the Milky Way than it is today, when light pollution is so common that many countries have identified "dark sky parks" from which to clearly see the stars. Away from the lights of a city, which make the night sky appear starless, the view is dramatically different, with a sweep of glowing embers illuminating the dark. Seeing stars from space, away from atmospheric interference and where it is even darker, may enhance the experience further. For an astronaut I call "Zack," witnessing this sight while in lunar orbit was transformative.

Meeting Zack

One late summer morning in 2017 I found myself driving to a restaurant in a medium-sized Midwestern town, both excited and deeply nervous. I had just recently published my first article in *The Space Review*, an experiment in reaching out to the space community and not just fellow anthropologists. To my surprise I was contacted soon after it ran by an Apollo-era astronaut. Zack wrote in an email that he had read my article with "great interest" and had "a very different philosophy about the place we have in the universe" than the astronauts who had been the subject of my article. Could I meet him to discuss the matter? He didn't have to ask twice.

Zack was charismatic, loved to laugh, and spoke loudly in the crowded restaurant about his journey to the Moon. It soon became clear exactly why he had wanted to talk to me. He met my eyes with a friendly yet intense stare, leaned forward, and said, "Looking at the universe out there from my vantage point, I began to realize that we don't know crap about anything. We really don't." His eyes seemed to look off into the distance. He continued, "At some points in my orbit around the Moon, I was in shadow from both the Earth and the Sun. It was a complete darkness. And all of a sudden, the star patterns out there became something that I was not ready for. So many stars I couldn't see *one*. It was just a sheet of light." I sat listening, starstruck. I was used to talking to pilgrims about miraculous healings and supernatural visions, but hearing someone who had been to the Moon (which, at the time of this writing, is still a group composed exclusively of white American men) describe what he saw from space was thrilling. Watching me carefully to make sure that I was following, he continued:

> I don't know if you call it spiritual or not, but when I saw the star field out there in a way that nobody else had ever seen, yeah, I had some pretty profound thoughts. Whether they were religious or not, I don't know. Let me tell you what I thought: There are so many stars out there. There's a pattern to the universe. There is a lifecycle of the universe. Carl Sagan would tell you that of all the stars out there, there are going to be a certain number of them that are like our Sun. There are going to be a certain number of *those*

> that have planetary systems. Of those planetary systems there are going be a certain number of *them* that are going to be like Earth. Of *those*, there are going to be a certain number that would sustain life, and of *those*, there are going to be a certain number that have life. They can never guess what the number is. If the number is 0.00001, that's still a positive number. You can't escape it. We are not unique in the universe.

Through his words, Zack had taken me to the Moon and allowed me to join him as he looked out into the Milky Way galaxy and saw no stars, only an unfathomable sheet of light. His experience of awe, an overwhelming sensation of realizing how little he knew in the face of something immense, beautiful, and baffling, sat me back in my seat. His awe was contagious and now I was feeling it myself, both electrified and struggling to comprehend his experience.

There exists a certain type of awe experienced only by a handful of people who've had the privilege of traveling to space. It's part of what I call the "ultraview effect," a cycle in which awe leads to humility and then to intense curiosity. Zack's experience of awe has only been shared by a few astronauts whose time in space allowed them to enter the deepest darkness, adjust visually, then gaze with the naked eye into the Milky Way galaxy—with no atmosphere to blunt the impact of seeing billions of stars simultaneously. Earthbound humans, even in the darkest areas, simply can't see these stars the way astronauts can under ideal conditions: as an immense and truly awesome intensification of the starry night sky, as Powys described above. The ultraview effect, currently limited to astronauts, is out of reach for most humans. However, the sensation of awe it evokes mirrors the universal human reaction to vast, unknowable mysteries, from the night sky to towering mountain ranges, deep oceans, or overwhelming monuments and architectural wonders.

Exploring Awe

The word "awesome," as it is used today, can mean that something is great or excellent, but the meaning of "awe" has a darker origin. Awe is "an emotion variously combining dread, veneration, and wonder that is inspired by authority or by the sacred or sublime," but its roots lie in the Old English *egsa,* meaning "fear, terror," and it is related to the Greek *áchos*, which "conveys distress or pain."[2] The contemporary meaning combines fear with pleasure, attraction with repulsion. It's uncomfortable wonder, agonizing and amazing all at once.

The concept of "awe" in English isn't universal. In Japanese, for instance, *ikei* conveys "respectful awe," while *ifu* indicates "fearful awe." Both blend fear and respect, with *ifu* carrying more fear than the English word "awe."[3] French has the concept of *admiration*, which expresses approval but not fear. To convey awe, francophones use phrases like *effroi mêlé de respect* ("fear mixed with respect") or *plaisir mêlé d'effroi* ("pleasure mixed with fear"), revealing that awe combines conflicting feelings.[4]

The Qur'an, written in seventh-century Arabic, often uses the word *khashya* to describe how humans should feel toward God. While *khashya* can be translated as "fear," English interpretations often use "awe" to represent a mix of reverence and humility in the face of divine greatness (Qur'an 2:74). This connection between awe and fear is apparent in members of religious groups who describe themselves as "God-fearing." In religious contexts, awe often acknowledges that some things are too vast or complex for human beings to fully comprehend, a feeling that can also emerge in deeply personal, secular experiences. One such moment of profound wonder unfolded for my family and me during a stargazing trip in Hawai'i.

In 2022 I visited the Big Island with my family, giving my son Luke, then aged fourteen, a chance to realize how much he didn't and *couldn't* understand about the universe. We had heard wonderful things about opportunities to stargaze on the extinct volcano of Mauna Kea, but the health of my mother-in-law, then eighty-five, made a mountain trip unwise. Stargazing at cold temperatures in a low-oxygen environment would have put her at risk. Fortunately, I found an opportunity to stargaze at sea level, at Hapuna Beach, on the rural northwestern coast of the Big Island. There, telescopes had been set up in a west-facing parking lot, and a crowd had gathered. Our eyes adjusted to the darkness, and the red lights around us gave us some ability to see. We were discouraged from using cell phones or flashlights.

The gorgeous sweep of the Milky Way gradually became visible over the horizon. I had seen glimpses of the Milky Way while crossing the desert between Southern California and Las Vegas, but this situation was altogether different. We spent more than an hour under a tremendous white curtain of light, with the organizers moving the focus of their telescopes from place to place as an array of cosmic lights slowly crossed the sky. Assisted by those powerful telescopes, we saw planets, stars, and entire galaxies from that beach, but even with the naked eye, the view was spectacular.

Luke hadn't seemed particularly interested in the outing ahead of time, but after a while he began to realize just what was going on above his head and out over the ocean. Something inside him began to shift. Sometimes he lay on his back on the hard concrete of the parking lot, staring up into space. He also enthusiastically made use of the telescopes, asking the organizers a lot of questions. He left some questions to me, asking how old the universe was and how long it took light to reach us from those stars. Until that evening, he hadn't truly absorbed the idea that the Sun was also a star, and the experience of looking at the Milky Way and understanding that every bit of light he saw was something like our own Sun impacted him deeply. Even in the dark I could see the gears of his mind moving as he processed new information, gradually realizing that there was so much above us he did not understand.

Seeing the Milky Way like that was a nearly universal part of human life before artificial lights began to crowd out the night sky. It was also a major source of awe, that blend of fear, admiration, reverence, and pleasure that is so widely recognized. The feeling of awe appears to be something essential to humanity. Stimulating that feeling is very important and can be done in numerous ways. There is something about the awe caused by looking at the stars, however, that seems to affect us profoundly. It's possible that the Milky Way has always been humanity's greatest source of awe. Is awe a physical thing that happens to our brains and bodies or is it a phenomenon of the mind? Is it built in, or something we must learn to experience?

Science and Awe

Berkeley psychologist Dacher Keltner defines awe very simply, writing, "Awe is the emotion we experience when we encounter vast mysteries that we don't understand."[5] Poets and scientists have also sought to describe this sensation, even though their

FIGURE 1. Stargazing at Hapuna Beach, illustrating the powerful experience of viewing the Milky Way from Earth. Illustration by Karl Tate.

perspectives often seem to be at odds. Walt Whitman, for example, in his 1855 poem "When I Heard the Learn'd Astronomer," shows how he, like my son Luke, found the experience of gazing up at the stars much more moving than listening to a dry lecture about their scientific properties. Whitman writes:

> When I heard the learn'd astronomer,
> When the proofs, the figures,
> were ranged in columns before me,
> When I was shown the charts and diagrams,
> to add, divide, and measure them,
> When I sitting heard the astronomer
> where he lectured with much applause
> in the lecture-room,
> How soon unaccountable I became tired and sick,
> Till rising and gliding out I wander'd off by myself,
> In the mystical moist night-air,
> and from time to time,
> Look'd up in perfect silence at the stars.[6]

For Whitman, the astronomer takes the mystery out of something he finds wondrous, trying to force his students to understand something that the poet would prefer to remain unknowable. This idea was put well by cartoonist Bill Watterson, whose character Calvin said in one comic strip, "That's the whole problem with science. You've got a bunch of empiricists trying to describe things of unimaginable wonder."[7] Whitman and Watterson suggest that the more you study something, the less awe you feel.

Whitman's "learn'd astronomer" may have taken the wonder out of the stars, but more soulful scientists, like Albert Einstein, Carl Sagan, and Rachel Carson have combined scientific inquiry with a love of the sublime and a recognition that some things are really unimaginable. Einstein, for instance, a physicist who combined mathematical skill and intuition to describe the theory of relativity, once said, "One cannot but be in awe when one contemplates the mysteries of eternity, of life, of the marvelous structure of reality. It is enough if one tries merely to comprehend a little of this mystery each day. Never lose a holy curiosity."[8] Einstein knew enough to recognize that a full understanding of the universe was beyond his grasp.

Sagan, an astronomer and science communicator, expressed a similar view when he wrote about trying to understand what he called "a little of this mystery." As a child, he had asked his mother about the stars and she recommended reading a library book. Book in hand, Sagan realized:

> It was in there. It was stunning. The answer was that the sun was a star, except very far away. The stars were suns; if you were close to them, they would look just like our sun. I tried to imagine how far away from the sun you'd have to be for it to be as dim as a star. Of course I didn't know the inverse square law of light propagation; I hadn't a ghost of a chance of figuring it out. But it was clear to me that you'd have to be very far away. Farther away, probably, than New Jersey. The dazzling idea of a universe vast beyond imagining swept over me. It has stayed with me ever since. . . . I sensed awe.[9]

Although Whitman and Watterson have suggested that increased scientific knowledge might dim a person's capacity for awe, that certainly wasn't Sagan's experience.

Carson, a marine biologist and conservationist, was another scientist who embraced awe. In a book aimed at parents, she wrote that good parenting included helping children understand the world scientifically. This would increase awe, not diminish it:

> If I had influence with the good fairy who is supposed to preside over the christening of all children, I should ask that her gift to each child in the world be a sense of wonder so indestructible that it would last throughout life. . . . If a child is to keep alive his inborn sense of wonder without any such gift from the fairies, he needs the companionship of at least one adult who can share it, rediscovering with him the joy, excitement and mystery of the world we live in.[10]

So scientists (at least some, and probably most) do experience awe and see it as an incredibly motivating emotion. Awe, it seems, makes scientists do science. It also shapes how philosophers, psychologists, and others make sense of humanity's shared experiences.[11]

Philosophical and Psychological Perspectives

Philosophers and other scholars have explored the idea of the sublime for centuries. The earliest known such work, *Peri hypsous* (*On the Sublime*), attributed to Longinus, dates back nearly two thousand years. Later, in the seventeenth century, Thomas Burnet's influential work *The Sacred Theory of the Earth* emphasized nature as a source of wonder by contrasting human limits with divine power, as described in the Bible.

In the eighteenth century, during the Enlightenment, philosopher Edmund Burke went further. His treatise *A Philosophical Enquiry into the Origin of Our Ideas of the Sublime and Beautiful*, written in 1757, differentiated between beauty, associated with harmony, tranquility, and pleasant feelings, and the sublime, linked to vastness, astonishment, and fear. According to Burke, beauty—found in gentle, delicate things like flowers or graceful buildings—was calming, while the sublime was overwhelming and often frightening, evoking "delightful horror."[12] An example he gave was the concept of infinity, which could create an unsettling yet thrilling sense of astonishment. While Burke called this experience the sublime, today it would be described as "awe."

German philosopher Immanuel Kant took up the idea of the sublime (once again, today's "awe") in his 1790 work *Critique of Judgment*. Where Burke believed the experience of the sublime was related to emotions like fear and astonishment, Kant thought it was more of an intellectual response. He argued that a sense of the sublime was actually produced when the mind came up against something so beyond the human senses that it was challenging to understand. If you try to understand it anyway, you experience the sublime. For Burke, sensing the sublime was more of a reflex, while for Kant, it only happened when humans took an active role, really trying to understand something confounding.[13]

Kant identified two main sources of the sublime. The first, the "mathematical sublime," occurs when humans attempt to grasp concepts like infinity or an endless universe, ideas our minds can't fully comprehend. The second, the "dynamical sublime," happens when encountering massive and overpowering natural phenomena such as hurricanes, mountains, or tsunamis, evoking a mixture of fear and admiration.[14] This may explain, in part, why tornado "chasers" seek out twisters instead of fleeing them.

Similar ideas about awe and the sublime also appear in non-Western traditions, like the Taoist reverence for nature and Indigenous cultural views that emphasize respect for the natural world, highlighting vastness and humility. Contemporary philosophers continue this exploration. Sandra Shapshay's "thick sublime" concept echoes Burke and Kant, exploring how the sublime affects aesthetic appreciation and ethical thought.[15] Philosopher Timothy Morton introduces "hyperobjects," massive entities like Earth or concepts such as climate change, which can be understood in part but never fully grasped. Morton emphasizes that the slow route to comprehending these hyperobjects, as one takes in new knowledge in pieces, can produce discomfort or even disgust.[16]

While the distinction between "the sublime" and "awe" isn't always clear, philosophers like Burke, Kant, and Morton tend to use the former, while psychologists typically use the latter.[17] Psychologists Yannick Joye and Jan Verpooten suggest awe began as a biological response—a "sensitivity to bigness."[18] Our evolutionary ancestors, they argue, needed to steer clear of large predators and dominant members of their own species; smaller animals that fearfully avoided large creatures had a survival advantage, potentially living longer and reproducing more. This tendency became reinforced and eventually encoded in our genes, laying the groundwork for awe as it is understood today.

Joye and Verpooten say that humans later leveraged this built-in fear response culturally. Impressively large objects could trigger a fear-and-respect response similar to that experienced when encountering dangerous animals. Builders of monumental, sacred architecture—like the pyramids of Giza in Egypt, the Great Mosque of Djenné in Mali, or Europe's Gothic cathedrals—capitalized on this reaction, granting their associated institutions authority and influence. The sheer size or "exceptional vastness" of these buildings could mentally "shake" viewers, making them more open to religious ideas. They illustrate this idea with Notre-Dame Cathedral in Paris, stating that its splendor and massive scale likely triggered awe in medieval people.[19] Awe, therefore, isn't merely a feeling that arises under the right circumstances—it may also transform, potentially increasing spiritual or religious inclinations in those who experience it.

Piercarlo Valdesolo and Jesse Graham, also psychologists, observed that subjects viewing awe-inspiring images (of mountains, canyons, outer space) had an increased tendency to attribute the creation of the objects to a supernatural figure, such as God.[20] Different images that evoked happiness, sadness, or other emotions didn't produce this effect. Valdesolo and Graham concluded that awe prompts people to seek explanations, often attributing these wonders to powerful supernatural beings. Keltner and his colleague Jonathan Haidt further found that awe increased openness to transformation and personal growth.[21] Awe-inspiring moments—whether viewing a vast canyon, appreciating intricate art or music, or witnessing a stunning sunset—encourage reflection and a readiness for change.

While philosophers typically use the term "the sublime," focusing on how awe blends fear, wonder, and pleasure arising from mentally confronting phenomena that may be difficult to understand, and psychologists favor the term "awe," emphasizing its emotional aspects, evolutionary origins, and role in fostering supernatural belief and personal transformation, anthropologists study how this universal emotion is culturally shaped and expressed.

Anthropological Perspectives

Anthropologists focus on what humans learn and understand through culture—that is, the knowledge, beliefs, values, and behaviors that groups hold in common and pass down to their children. Language and religion are part of culture. So are morals, approaches to humor, gender roles and expectations, what to eat, and how to dress. Anthropological research on awe reveals it to be another part of human life that is cultural. For that reason, anthropologists studying awe focus less on individuals and more on societies. Awe is understood as something shaped by each cultural group through religious rituals, social structure, and shared emotions.

Anthropologist William Murphy used this understanding in his research among the Mende people of the West African nation of Sierra Leone. The Mende practice agriculture, farming rice and other crops like yams and cocoa. Murphy became interested in a Mende concept called *kabande*. Murphy describes *kabande* as being similar to what is called "awe" in English, explaining that it's used to describe something greater than oneself, something powerful, mysterious, and beautiful. *Kabande* can be used to describe feelings created by a person's political power, the sublime movements of masked dancers, or religious and spiritual forces. Its similarity to the idea of awe helps support the idea that awe or similar feelings may be found across humankind.[22] While Murphy examines awe through the spiritual and social lens of *kabande* in Sierra Leone, other scholars, like Matthew Rampley, explore how awe is evoked by human-made marvels, such as architecture and technology.

In his article "The Ethnographic Sublime," Rampley looks specifically at awe inspired by technology. While people usually associate awe with natural wonders, Rampley explains that impressive human creations—like huge bridges, city skylines, or space missions—can also create feelings of awe.[23] He references historian

David Nye's study of tourism at Niagara Falls; in *American Technological Sublime*, Nye shows that awe isn't always immediate but can build gradually, shaped by what visitors already know or expect. For many tourists, awe at Niagara Falls includes amazement at the technology needed to control its power.[24]

Rampley's ideas align with Joye and Verpooten, who note that monumental structures like cathedrals and pyramids often inspire awe. Yet, similar to Murphy, Rampley emphasizes that some parts of awe come from what we learn from our cultures rather than from biology alone. According to Rampley, what people find awe-inspiring depends greatly on their cultural practices and language, leading to differences between societies.[25]

Rampley also argues, however, that while the specific things that inspire awe vary across cultures, the basic human ability to feel it is likely a result of evolution. This connects back to psychological approaches. Looking across different societies, Rampley notes that powerful displays—whether natural, human-made, or political—universally inspire awe. The exact ways awe is expressed and the things that trigger it, though, depend heavily on cultural context. Experiences described by Burke and Kant may differ significantly from those from cultures like the Mende or various Indigenous traditions. Therefore, awe is both a universal human emotion and uniquely shaped by culture.[26]

In my own work, I've studied how our built-in ability to feel awe interacts with cultural differences, especially through pilgrimage, which can be defined as travel for religious reasons. For more than thirty years I've researched Rocamadour, a Catholic pilgrimage site in France. Rocamadour is known for its "black Madonna," a centuries-old statue of the Virgin Mary that is literally black in color, probably due to tarnished silver and years of candle and incense smoke. The Madonna is displayed in a small chapel halfway up a 150-meter cliff, next to a basilica, both structures seeming to defy gravity. Three of the chapel walls are typical medieval constructions, but the fourth wall is the actual cliff, creating a space that feels both indoors and outdoors.[27]

Catholic pilgrims visit the site, but it's also important to French tourists, most of whom aren't religious at all. Visitors also include pilgrims practicing earth-based religions like neopaganism and participants in what is sometimes still called the "New Age" movement. I spent many months living at Rocamadour, talking to both its residents and visitors. I was struck by how the religious and nonreligious, as well as people following both traditional and nontraditional spiritual paths, all talked about "feeling something" at the site. I argue that there's something about interacting with the cliff, whether climbing it via paths or staircases, or looking up at it from underneath or down from its height into the chasm below, that evokes awe. Once someone feels that awe, their particular religious beliefs (or lack thereof) shape how they understand their experience. One person may describe the feeling as the power of God or "the goddess," while others may attribute unusual sensations to underground electrical currents or the presence of microwaves. Once again, a shared experience is still shaped and molded by learned knowledge, beliefs, values, and behavior.[28]

Awe is a powerful emotion but less common than emotions like joy, anger, or sadness. Some people may rarely experience awe, but when they do, it can deeply affect their lives. For others, awe is something they actively pursue. Just as Rocamadour's cliff creates awe by challenging visitors' perceptions of scale and place, the ultraview effect takes this feeling to another level entirely in the vast, alien environment of space. To understand why certain experiences in space trigger awe, it's important to recognize how strange and biologically unfamiliar outer space is to humans. Our species is fundamentally connected to Earth, shaped by it, and linked to it no matter how far we go.

Terrestrial Creatures

As much as human cultures are drawn to the sky and the space beyond it, human beings, from both religious and scientific perspectives, are earthbound creatures. Years of research in a range of disciplines show that life here has adapted well to the world's environments. A diverse community of simple life forms gradually gave rise to one-celled organisms, which themselves evolved and transformed, very slowly, leading to a vast array of multicellular plants, fungi, and animals.[29] These plants, fungi, and animals continued to change and take the forms of new species, many of these replacing the old ones and joining the life forms that exist today. We can think of evolution and speciation (the process by which one group splits into separate groups that can no longer interbreed) as transformations of the first forms of life on Earth. This means that all growing things on Earth are essentially the same kind of life, with new forms simply reflecting adaptations to new conditions. The genetic coding in our DNA strands reveals that human beings, termites, maitake mushrooms, penguins, pine trees, and sharks are all basically the same *type* of thing, just adjusted to different circumstances.

All known life forms share a common biological origin, but they also share the Earth as their habitat. Like other living things, human beings are physically composed of the molecules and minerals found on Earth, building blocks that underscore whence humans come. In 2008, journalist Natalie Angier interviewed Dr. Andrew Schafer (then president of the American Society of Hematology), and wrote:

> Blood can also be thought of as a private ocean, a recapitulation of what life was like for all the years we spent drifting as microscopic, single-celled organisms, "taking up nutrients from sea water and then eliminating waste products back into sea water," Dr. Schafer said. Not only is blood mostly water, but the watery portion of blood, the plasma, has a concentration of salt and other ions that is remarkably similar to sea water.[30]

Humans and other life forms, already similar to one another, are also similar to the Earth itself. For example, most of us serve as habitats to other life forms, most of them microscopic. In his wonderful book *I Contain Multitudes: The Microbes within Us and a Grander View of Life*, science writer Ed Yong explains how diverse living locations, such as the human gastrointestinal tract, the mucus layer of reef-building coral, and the antennae and cocoons of wasps, all host and even rely on bacteria for things like digesting food and healing injuries.[31]

It's not just the Earth that influences our biology. All of the energy on the planet—heat and light, for instance—ultimately comes from the Sun. The Moon, too, appears to have an impact on humans and other life forms. Scientific studies reveal, for example, that the level of light produced by the Moon at different phases has an influence on the behavior of certain primates. Latin American night monkeys, also known as owl monkeys, are much more active during nights with full moons than with new moons, while Indonesian spectral tarsiers, whose eyes are adapted to low-light conditions, are actually *less* active during nights with full moons.[32] Sea creatures also experience behavioral and biological changes in response to high and low tides.[33]

Humans need the right amount of air pressure, gravity, and oxygen levels to live. If we go to the bottom of the ocean outside of a vehicle or pressure suit, our bodies will be crushed. At certain altitudes, such as those found at the top of Mount Everest, there's not enough oxygen for a human to survive. We evolved under conditions that proliferate on planet Earth. I can find adequate oxygen in the jungles of Amazonia, the beaches of Japan, and the deserts of California. I can move up and down different latitudes or across different longitudes and still have enough pressure to keep my chest from collapsing or my gases from evaporating out of my body. It's not so much that humans are on a "Goldilocks planet" where everything is "just right," but that our evolutionary ancestors and countless DNA cousin species developed and thrived under the same circumstances, more or less, in which we still find ourselves.[34]

During different parts of the year, Earth's distance from the Sun varies by more than three million miles. Even if Earth had settled into an orbit significantly closer to or farther from the Sun, life might still have evolved—though it would look and function very differently. Think of it this way: If I pour water into a bowl, it takes on a bowl's shape; if I pour it into a champagne flute, it becomes champagne flute-shaped. Similarly, life developed specifically in response to Earth's conditions, and all its forms reflect our planet's unique influence.

Recognizing that our bodies and minds have been formed by Earth's precise conditions—shaped by oceans, sunlight, moonlight, and gravity—makes it clear why humans experience awe when confronting environments that diverge dramatically from our terrestrial norms. Awe often arises precisely because it emphasizes the contrast between expected conditions and phenomena that challenge or exceed those familiar boundaries.

The important place of awe in human life seems very clear when we look at its ubiquity, appearing in many cultures and being studied by multiple disciplines. While awe can certainly have religious or spiritual dimensions, it's not necessary to believe in anything supernatural to experience awe—all it takes is a recognition of phenomena that are just too enormous or vast for us to ever truly understand. Contemplating the size of a blue whale, the scale of the Himalayas, or the impact of a black hole on its galaxy can produce awe in anyone who opens their imagination to the reality of those things.

Even as we contemplate the mysteries of space, the feeling of awe reminds us that we remain earthbound creatures, grown from the fabric of this planet and subject to its influence and rules. Because humans evolved in circumstances marked by a specific amount of gravity, a certain percentage of oxygen, and the protective shield of a magnetic field, our survival depends entirely on these conditions. Human bodies reflect the unique features of Earth.

The exploration of space—where humans are the proverbial fish out of water—reinforces how dependent humans are on our planet. This awareness evokes both the fear of danger and the thrill of discovery. The human capacity for awe drives some of us toward learning about the unfamiliar vastness beyond our planet, challenging the limits of how far one can travel and still survive. Yet, as much as some people seek to escape Earth's boundaries, human bodies and minds remain fundamentally shaped by the planet we evolved from—a tension that has defined humanity's journey from ground to sky.

2 *Ground to Sky*

The Human Journey to Space

In her memoir *High, Wide and Frightened*, Louise Thaden, a pioneering aviator of the 1920s and 1930s who fought for recognition in a male-dominated profession, writes, "Flying is the only real freedom we are privileged to possess." This thought comes as she builds sandcastles on the beach with her young son, distracted by birds flying overhead and yearning to return to the skies herself.[1] While Thaden was temporarily grounded by motherhood, her longing reflects a broader human desire to ascend—toward the sky, stars, or supernatural realms—a recurring theme across societies.

Our minds and bodies evolved to function best within an Earth-based context. Culture, on the other hand, is learned rather than a direct outcome of biological evolution. Cultural innovations vary widely, from incremental improvements to radical shifts. For example, early humans initially ate food raw, but the invention of cooking not only made food easier to digest but also rendered previously inedible items nourishing and appealing.[2] Other innovations required even greater imaginative leaps, pushing humans into entirely new realms: inventing reading and writing systems; designing furniture; fashioning protective gear, like shoes and helmets;

creating movies, sporting equipment, and card games; and developing mechanical modes of transportation, like bicycles, automobiles, and aircraft. Human adaptability has consistently transformed the unimaginable into the commonplace. Communication technology, for instance, once limited primarily to affluent populations in the Global North due to its infrastructure requirements, quickly spread worldwide through mobile and satellite phone networks—a phenomenon known as "leapfrogging."[3]

Spacecraft would not exist without airplanes, and the leap from walking to flying was far more profound than the step from gliders to rockets. Although ancient myths around the world describe magical flight or travel on winged animals (such as Pegasus or Garuda),[4] practical human flight emerged only through technological innovations, beginning in the eighteenth century in Europe. The Montgolfier brothers' hot air balloon was the first successful aircraft,[5] soon followed by zeppelins, blimps, and eventually airplanes. Still, all these early aircraft operated within Earth's life-rich troposphere. Humans reached the stratosphere for the first time in 1931, when Auguste Piccard and Charles Kipfer ascended in a balloon to altitudes inhabited only by a few hardy microorganisms.[6] True departure from Earth came three decades later, with Soviet cosmonaut Yuri Gagarin's groundbreaking spaceflight in 1961.[7]

Each new approach to flight introduced novel experiences, perspectives, and a sense of breaking limits. As Thaden observed, flying symbolizes freedom and transcendence. During this relatively short period when humans have ventured into places previously unreachable, our species has repeatedly encountered the powerful mixture of fear, excitement, pleasure, wonder, and dread that we call awe.

Taking to the Air

Humans had to master powered flight through air before attempting the even more challenging leap into space. The first credited instance of human gliding—flight without engine-powered thrust—dates back to Abbas ibn Firnas, a ninth-century inventor of Arabic descent, born in what is now Spain. Using silk and eagle feathers, he tested principles of flight by launching a glider from a high point at Jabal al-'Arus, in the Rusafa area near Córdoba. According to historical records, he stayed airborne for about ten minutes before crashing and injuring his back. His pioneering experiments later inspired Leonardo da Vinci's Renaissance-era ornithopter.[8]

Most aviation historians recognize Clément Ader, born in 1841 in Muret, France, as the creator of the first airplane. Ader, who coined the French word *avion* (airplane), derived from the Latin *avis* (bird), developed innovative, steam-powered, batlike flying machines. On October 9, 1890, he took off in his *Ader Éole* and left the ground briefly but never achieved controlled flight.[9]

The most commonly recognized first instance of controlled, powered flight occurred in 1903 at Kitty Hawk, North Carolina, with Wilbur and Orville Wright's famous Wright Flyer. Some, however, credit Brazilian Alberto Santos-Dumont with that honor, due in part to limited press coverage of the Wright brothers' initial flights in Europe.[10] This perception persisted for multiple reasons. For instance, Santos-Dumont publicly demonstrated successful flights in 1906, taking off and landing independently on Paris's Bagatelle grounds, while the Wright brothers, before their 1908 demonstrations in Le Mans, had sometimes relied on catapults, raising public skepticism about their claims.[11]

FIGURE 2. Abbas ibn Firnas, an early and often overlooked pioneer of aviation, testing his glider in ninth-century al-Andalus. Illustration by Karl Tate.

Even as flying became increasingly common during the technological boom of the early twentieth century, human flight still felt astonishing and miraculous. Joseph J. Corn, author of *The Winged Gospel: America's Romance with Aviation, 1900–1950*, describes enormous, amazed crowds at aviation exhibitions. According to

Corn, observers described flight with mystical language, using terms like "miraculous," "inhuman," or even "occult."[12] Corn also references sources from the 1920s, indicating that many Americans at the time believed that airplanes could literally fly people to heaven; one elderly Black woman approached Charles Lindbergh and offered to pay him to take her on a one-way trip to the pearly gates.[13] This example shows just how wondrous and awe-inspiring human flight was thought to be during those early days of aviation.

Warbirds and Rockets

The idea that airplanes could take a person to heaven was quickly replaced by the realization that they could be used as hellish tools of warfare. Even during Ader's time, nations were interested in the military applications of flying machines, and by 1914, the use of airplanes in battle had become common.[14] During World War II, warplanes were joined by rockets, like Germany's V-2 and America's M8.[15] These new inventions could both devastate and elevate, inspiring both awe and fear.

Although rockets and other projectiles had been used in warfare for centuries, the first liquid-fueled rocket was invented by Robert H. Goddard in 1926. Goddard's passion for space travel reportedly started in childhood, when he would climb a cherry tree, look up to the sky, and imagine the possibility of reaching Mars.[16] Despite this interest in interplanetary travel, much of Goddard's work focused on military applications. During World War I he collaborated with the U.S. Army Ordnance Department on rocket launchers and, before his death in 1945, had made substantial progress stabilizing rocket flight through gyro control apparatuses.[17] His work proved influential in World War II, when rockets and guided missiles were increasingly used to raze cities and sink battleships—purposes far removed from Goddard's childhood dreams of space travel.

One significant rocket engineer influenced by Goddard was Wernher von Braun, who oversaw Nazi Germany's development of the V-2 rocket. The V-2 was built under horrific circumstances, using forced labor from concentration camps. Nearly 20,000 prisoners at the Mittelbau-Dora production site died from exhaustion, starvation, and abusive conditions.[18] Despite his dark history, Von Braun was brought to the United States along with approximately 350 other German scientists through an initiative called Project Paperclip, designed to leverage their scientific expertise at the start of the Cold War and gain an advantage over the Soviet Union. The devastating use of nuclear weapons in World War II had inspired a type of awe far more connected to terror than appreciation, and nations worldwide hoped the kind of horrifying destruction witnessed during the war would never occur again.

Germany had been working on nuclear capabilities throughout the conflict, meaning that German scientists held valuable information potentially of interest to the Soviet Union. Project Paperclip permitted scientists directly involved in Nazi research and military institutions to evade trials and escape justice for war crimes by agreeing to work for the United States, their former enemy.[19] Anticipating the scandal this arrangement would cause, the U.S. government suppressed knowledge of the program from the American public. Only in the early 1970s were its full implications revealed to an outraged citizenry.[20]

While the U.S. and Soviet Union competed for military capability, much of the Cold War rivalry was symbolic. A hot war could destroy the world, so the superpowers turned to feats of strength that avoided direct combat, including the exploration and use of outer space. Scientists like Goddard had long dreamed of sending humans into the cosmos, and World War II—despite its horrific destruction—accelerated technological advances that helped make spaceflight possible. Innovations in rocket propulsion, guidance, and materials later became valuable to space exploration, but they also fueled devastating conflicts in Korea and Vietnam, where entire regions were destroyed. For the public, however, it was more pleasant to imagine rockets as vessels for brave explorers than as weapons that had once devastated cities. To oversimplify a bit, a rocket's reputation could be partially rehabilitated by replacing its payload: take out a warhead and insert an astronaut. The military's interest in rockets continued, and the threat of nuclear annihilation remained very real, yet the growing space program offered a hope that technologies born in war might be redirected toward less overtly violent purposes.

A View from Space

Peaceful ambitions, however, were countered by the tension that existed between the Soviet Union and the United States in the early days of each nation's space program. Part of this strain came from differences in creed. As the Soviets followed the ideas of Karl Marx, seeing religion as a tool that czars had used to control and suppress the people, the United States countered by promoting an image of Americans as strong religious believers blessed by God.[21] As a result, many Americans, including quite a few astronauts, saw the space race as a contest in which national values—whether atheistic or Christian—were being tested in the stars. The successes or failures of the cosmonauts and astronauts were often seen as proof of their countries' underlying belief systems.

You can see this view in the words of "James," an Apollo-era space technician who explained that it was the religious faith of Americans that allowed them to get to the Moon first. Referring to the turbulent late 1960s, James said, "You know with so many different bad things going on at that time, [when] you can do something like that achievement, I think it's a higher power that says, 'Yeah, let's go do this,' where other countries didn't do it." For James, who deeply valued his work on Apollo, American success in space was nothing short of a gift from God.

Americans were first to the Moon, but they weren't the first in space. That honor went to Soviet cosmonaut Yuri Gagarin, who was launched into orbit from Baikonur, Kazakhstan, on April 12, 1961.[22] The strong atheistic ideology of the Soviet Union at the time meant that Gagarin's flight wasn't prayed over or blessed by a Russian Orthodox priest, ceremonies that often take place before contemporary Russian space missions.[23] Although the U.S.S.R. officially adhered to atheism, elements of ritual and tradition were connected to that flight. For instance, after Gagarin's successful launch and return, cosmonauts began replicating his actions, driven by what anthropologists sometimes call, perhaps problematically, "magical thinking."[24] The belief that following Gagarin's steps could protect them from disaster led to the tradition of stopping their transport to the launch pad to urinate on the back tire—a ritual that started when Gagarin, likely feeling the nerves of anticipation, had to relieve himself before his historic flight.[25]

Another mystical influence, one that was certainly not made known to the Soviet public or international audience, was the philosophical movement known as Russian cosmism, which was most popular in the late nineteenth and early twentieth centuries. The philosophy blended Western and Eastern religious ideas with more esoteric belief systems. One of the leaders of Russian cosmism was Nikolai Fedorovich Fedorov. He believed that the universe was conscious in a chaotic way and that humankind was destined to go into space to bring order to this chaos. He also believed that "scientists of the future would be able to resurrect the dead by gathering and reassembling their dispersed atoms," which also required a human presence in space.[26]

His compatriot, cosmist and rocket scientist Konstantin Tsiolkovsky, was likely one of the first to describe the Earth as a "cradle." In this view, the planet was never meant to be the permanent home of humanity but instead something more like a nest from which our matured species would emerge.[27] Remarkably, ideas similar to Tsiolkovsky's notion can be found in the American space community even now. During a 2024 talk by Apollo 9 astronaut Russell "Rusty" Schweickart upon his receipt of the Explorers Club's "Legendary Explorers Award," he discussed an idea he called "cosmic birth," which views the human exploration of space as something akin to a baby being born. Just as an infant growing in the womb reaches a size where it takes too much from its mother and must leave her body, Schweickart explained, humanity is reaching a similar stage with our planet. He said, "We're expecting too much; we're demanding too much from the planet to provide for the growth that we're involved in intrinsically and producing too much waste." He then pointed out that it's only when a baby is born that it recognizes its mother as separate from itself and is able to love the mother as a distinct individual, stating, "Before birth, it's all one way; mom's doing everything. But after birth, we begin to recognize mom . . . and the love flows two ways." It was fascinating to hear an idea reminiscent of Tsiolkovsky from an American who had actually flown in space.

This kind of harmony in Russian and American ideas was less common during the Cold War, when the space programs of both countries were locked in a technological race. By 1965, it seemed that the Americans were losing to the Soviets in each new milestone of space exploration. As I have noted, Gagarin was the first human in space, as well as the first person to go into orbit. A Soviet was also first to perform a space walk, also known as an extravehicular activity, or EVA. On March 18, 1965, as part of his Voskhod 2 flight, Alexei A. Leonov became the first human to float in space outside a spacecraft. His experience was challenging, and reports came back that Leonov had struggled to regain entry into his capsule and later experienced medical issues.[28] Despite these concerns, Leonov's overall success provided extra motivation for the United States to conduct its own space walk. The crew of Gemini IV, specifically its pilot, astronaut Ed White, was selected for the honor.

On June 3, 1965, White opened the hatch of the Gemini IV craft and then moved weightlessly in space, attached to the vehicle by a tether. He also had a "zip gun" that emitted jets of oxygen, allowing him to maneuver in different directions. During this space walk, White was in radio contact with James McDivitt, the mission commander, who remained inside the capsule, and Christopher Kraft, the on-ground flight director. As Barton C. Hacker and James M. Grimwood wrote in their classic *On the Shoulders of Titans: A History of Project Gemini*:

> White told McDivitt and the world how beautiful it all was, of the pictures he was taking, and how well he was feeling—no vertigo or disorientation whatever. And when McDivitt had to tell him it was time to come back inside, Mission Control and the whole world

heard him sigh, "It's the saddest moment of my life." While he was floating freely, White had paid no attention to the time; and, since they were on the internal spacecraft communications link, Flight Control could not break in on them. Finally, after 15 minutes 40 seconds, McDivitt broke off to ask the ground if they wanted anything. "Yes," Kraft chuckled, "Tell him to get back in."[29]

The experience that Ed White had in space, one that he clearly wanted to continue, is a strong example of the ways in which space exploration can lead to incredible feelings of awe. White's hesitancy to return—his "saddest moment"—showcases how space exploration can lead to feelings of wonder and liberation, a sense of being unhampered by the confines of the Earth. Experiencing microgravity inside a space vehicle—especially one the size of a small car—is awe-inspiring in its own right. But to hover in nothingness with the reality of your own planet hanging next to you, its true size overwhelming, is something entirely different.

While White's EVA was a first in American spaceflight, a similar, and perhaps even deeper version was experienced by Michael Collins during his Gemini 10 space walk. Collins, who traveled into space with fellow astronaut John Young, was selected to conduct an EVA on July 20, 1966. He was tasked with dismantling a meteorite detector on the exterior of the spacecraft. Although he had some difficulty getting out of the capsule—Collins had inadvertently left his left shoulder harness connected and was restrained from exiting the vehicle until he detached it—once he was drifting gently in space he had a sublime experience, detailed in his memoir *Carrying the Fire*. Collins describes a feeling of surprise from a sense that he's gone "outside," then describes the awe evoked by having his visual field dramatically expanded from what he could see through the capsule window. He writes, "My God, the stars are everywhere: above me on all sides, even below me somewhat, down there next to that obscure horizon. . . . This is no simulation, this is the best view of the universe that a human has ever had."[30] It's evident from Collins's beautiful prose that he was swept up in this moment, feeling elation and amazement in equal measure.

A little more than two years later, in December 1968, the crew of Apollo 8 traveled into lunar orbit. During their journey, astronauts Jim Lovell, Frank Borman, and Bill Anders had two awe-evoking visions that had never before been experienced by humankind. The first was seeing the Moon up close from lunar orbit, and the second was observing the Earth from the vantage point of the Moon, nearly 235,000 miles from our world. Both experiences left the crew amazed. In a telecast broadcast to Earth on December 24, 1968, Frank Borman said of the Moon, "My own impression is that it's a vast, lonely forbidding type existence, great expanse of nothing . . . and it certainly would not appear to be a very inviting place to live or work." Borman acknowledged that each member of the crew had his own impression; proving him right was Jim Lovell, who was more positively impressed and said, "The vast loneliness up here of the Moon is awe inspiring and it makes you realize just what you have back there on Earth. The Earth from here is a grand [oasis] to the big vastness of space."[31]

While photographs taken by airplanes and satellites give us a glimpse of the Earth from above, imagining what the Moon looks like up close or what it would be like to see the Earth as a small blue orb from lunar orbit requires a greater leap. Even the awe of soaring over the world's geographic marvels like the Andes or the Grand Canyon can only hint at the wonder of seeing the Moon's Montes Apenninus with the Earth shining brightly overhead.

The Overview Effect

The "overview effect," a term coined and a concept investigated by journalist Frank White (no relation to Ed White), describes what happens to some astronauts when they are in a position to see the Earth from a spacecraft at some distance from the planet, whether in low Earth orbit or from further away.[32] This perspective was only available to us after human spaceflight began in the 1960s, and its origins as a popular concept seem to have roots in the 1968 Apollo 8 mission.

When the Apollo 8 crew flew toward the Moon and were able to see the Earth from a distance, they sent back photos that created quite a stir. While Apollo 11 is arguably the best-known NASA mission, with its successful Moon landing and Neil Armstrong's famous first steps, and Apollo 13 is famous because of how its crew narrowly escaped disaster and inspired a successful Hollywood movie, many space aficionados hold Apollo 8 close to their heart. Apollo 8 wasn't as dramatic as 11 or 13, but it was the first time humans flew to another celestial body. When Anders, Borman, and Lovell orbited the Moon, they proved that it was possible to actually reach it, bringing John F. Kennedy's promise to land a man there that much closer to achievement.

On the morning of December 24, 1968, during Apollo 8's fourth orbit of the Moon, the command module, CSM-103, changed direction and the crew was able to see the beautiful sight of the Earth rising above the surface of the Moon. The spacecraft's audio track captured the moment when Bill Anders spotted our planet, exclaiming, "Oh my God, look at that picture over there! There's the Earth coming up. Wow, is that pretty!"[33] Proving that awe doesn't come easily to everyone (or that he didn't quite see what

Anders was looking at), Borman answered, "Hey, don't take that. It's not scheduled." Anders requested a camera from Lovell anyway, specifying his intent to take a "color exterior" shot. After a bit of fumbling, Anders was able to snap the picture that has come to be known as "Earthrise."[34] Andrew Chaikin interviewed Bill Anders, who talked about that moment. Chaikin writes:

> "That was the most beautiful thing I'd ever seen," said Anders. "Totally unanticipated. Because we were being trained to *go to the Moon*. . . . It wasn't 'going to the Moon and looking back at the Earth.' I never even thought about that!" Seeing the Earthrise, Anders told me, changed his view of the mission in real time. "In lunar orbit, it occurred to me that, here we are, all the way up there at the Moon, and we're studying this thing, and it's really the Earth as seen from the Moon that's the most interesting aspect of this flight."[35]

The "Earthrise" photo, beamed back to the planet it depicted, was a sensation. Although photos of the Earth and Moon taken by uncrewed probes had been shared previously, the Apollo 8 images, perhaps because human vision and intention were involved (anyone who had snapped images on a vacation knew the process, so it was easy to relate), seemed to strike a different chord. These photographs ushered in a new era of eco-consciousness and a stronger sense of the Earth as an isolated and beautiful spacecraft, keeping all of us going with its organic life-support system.

As this new vision, essentially a new understanding of the Earth, was spreading, journalist White began to notice a few things. He was first inspired to think about what the Earth looked like from above during a flight on an airplane. Looking down, White realized that a physically elevated person could understand more about the

FIGURE 3. The iconic "Earthrise" photograph taken by astronaut Bill Anders during the Apollo 8 mission, considered by many to have launched the overview effect. Courtesy of NASA.

larger picture than someone on the ground. Two cars might be approaching each other from around a corner, he wrote, but since they weren't in each other's line of sight, the drivers wouldn't know that. An observer from above, however, would have an "overview," seeing a relationship and a reality beyond what could be perceived by the drivers on the ground. Astronauts on a space station or spacecraft, White reasoned, would have a similar take on the world based on their heightened point of observation.[36]

White was onto something. As he interviewed astronauts it became clear that for most, seeing the Earth from space was one of the true highlights of spaceflight, an amazing path to awe unavailable to those of us with more ground-based perspectives. Astronauts spoke of the shock of seeing the thin blue line of the planet's atmosphere, the long stretches of ground unbroken by borders (which seemed quite arbitrary viewed from space), and the sudden and unsettling realization that the Earth really is a giant, weather-streaked orb rotating in nothingness.

In the fourth edition of White's book *The Overview Effect: Space Exploration and Human Evolution*, he includes an interview with Richard Garriott de Cayeux, a spaceflight participant who paid to fly on a Roscosmos mission in 2008. Garriott describes the moment: "My sense of the scale of the Earth suddenly collapsed," he recalled, "and I had an extremely physical, visceral, strong reaction at that moment, like 'I get it, I get the Earth.' . . . I now feel like I understand it in a way that I never could have prior to that moment, and to me it went from being unambiguously and unknowably large to being something that was clearly a finite, small, and delicate oasis."[37] White coined the term "overview effect" to capture this phenomenon that many astronauts had described. After Apollo 8, humanity seemed ready for such a concept and the phrase was soon in use, both on Earth and in space. Our species had an amazing new way of seeing our world, and the new phrase served a real purpose, marking this vantage point as special with a precise and enlightening term.

In my research with astronauts, it became clear that the overview effect is known and expected, although not universally held. The responses vary in ways that reflect my interviewees' diverse backgrounds, beliefs, and expectations.[38] One astronaut I call

"Alan," for instance, had expected to be wowed by his view of the Earth. Instead, he had been underwhelmed. He said:

> The first time I looked out at the Earth from space, I intentionally paused and collected myself and meditated a little bit to clear my head before I opened my eyes and looked out the window for the first time. And I didn't really feel anything. It's kind of a letdown. There was nothing. And maybe it's because I'm not a spiritual person, that's quite possible. . . . It was a beautiful sight and a unique vantage point, but there was nothing about it that I felt in any way unlocked any kind of philosophical mysteries or spiritual mysteries. . . . The one thing that people ask me is, did anything change? Did my spatial experience change me in any way? I think it just gave me a better appreciation for this planet, a bigger concern about climate change and other ways that we are trashing the place. But as far as that overview effect and . . . getting a sense that we are a common humanity and . . . we all share a common home and the things that unite human beings . . . are greater than the things that divide us? I think the . . . reason that I didn't come away with some kind of sudden, strong, compelling feeling in that regard is that I think I knew that before I left.

As for "Paul," a shuttle astronaut who flew in the 1980s, seeing the planet from above wasn't as impressive as floating in microgravity or the sleek functionality of the shuttle itself. Paul's overview of the earth was more depressing than enlightening. The fragility of the planet didn't inspire him but drove home that human beings are fully capable of destroying themselves. He gave our species five hundred to one thousand years at most before we wipe ourselves out.

While Alan seems not to have experienced awe, Paul's perspective brings the elements of dread and discomfort that can contribute to it. The overview effect, seemingly, isn't a purely positive experience and may also bring difficult emotions to the surface. This is apparent in the experience of William Shatner, who went to space via a private Blue Origin rocket in October 2021. While NASA astronauts have learned to process their experiences discreetly, often waiting until retirement to express more emotional responses, those outside the agency, like Shatner, can offer immediate, more expressive accounts that sometimes hint at darker aspects of the overview effect.

The morning of his flight, I sat at my office computer and watched spellbound as Shatner's mission, NS-18, landed. The crew of new spacefarers slowly left their capsule to stand again in the West Texas desert. Jeff Bezos, who runs both Amazon and Blue Origin and had recently flown himself, was there to greet them. He approached Shatner, then ninety years old, who was visibly moved. Bezos looked enthusiastic, clearly expecting a positive review of the event from the man who had played Captain James T. Kirk on the original *Star Trek* television series. Someone who had embodied the brave conquest of space to generations of television viewers had finally left the planet for real—what did he think?

Looking increasingly uncomfortable, Bezos listened to Shatner describe space as "ugly" and connect his spaceflight to feelings of death and dying. Shatner, caught up in the moment and live streamed around the world, told him:

> To see the blue color go, whip by, and now you're staring into blackness. That's the thing, the covering of blue . . . this comforter of blue that we have around us. We say, "Oh, that's blue sky," and

> then suddenly you shoot through it all of a sudden . . . and you're looking into blackness, into black ugliness. And you look down. There's the blue down there, and the black up there, and it's just there is mother, and Earth, and comfort. And there's . . . Is there death? I don't know. Was that death? Is that the way death is?

A few moments later, after discussing the rapid rise through and out of the blue sky he had experienced during his journey, Shatner again said, "And you're in death. . . . This is life and that's death. And it's in an instant, you go, 'Whoa! That's death.' That's what I saw."[39]

NASA astronauts have been rigorously trained, many of them in the military, and are generally reluctant to say anything controversial, at least not when there's still a chance of participating in future space missions (retired astronauts tend to be a *lot* more outspoken). William Shatner, however, was an actor, an artist—someone skilled at conveying emotion. Rather than holding back his feelings or speaking carefully, he instead shared his unfiltered thoughts, immediately after he had returned to Earth. So the Earth was life and space was death.

He conveyed awe impressively, demonstrating wonder, dread, and fear. He also expressed certain aspects of the overview effect very clearly, especially a new awareness of our planet as a real, physical object, something he had seen from outside. This external view conveys knowledge more forcefully than reading something in a book or seeing a documentary. Shatner understood the concrete characteristics of our home world in a way he hadn't been capable of comprehending previously. He explained his feelings: "What I would love to do is to communicate as much as possible the jeopardy, the moment you see . . . the vulnerability of everything, it's so small. This air which is keeping us alive is thinner than your skin. It's a sliver. It's immeasurably small when you think in terms of the universe. It's negligible, this air. Mars doesn't have it. . . . It's so thin."

Some people who have been to space don't experience the overview effect at all, taking in an external view of the Earth without responding with wonder, dread, or delight. Others have experienced a darker form of awe, an amazement paired with foreboding. Most, however, seem to relish the experience as one of astonishment and admiration. Space Shuttle–era astronaut "Beverly" said she didn't think the benefits of seeing the Earth from space were necessarily limited to those who actually traveled off-planet and hoped her children would reach the understanding that she had attained without needing to be astronauts themselves. Despite this wish, however, she described her own vision of the Earth in a way that showed actually being there made a huge difference. She told me, "It's this experience that puts that reality in your face in a way that you can't deny. You're like, 'Oh my gosh, that is a planet! Look at that, everybody I know is down there. It's really beautiful.' All those things. 'Look at that little super thin wafer line that's all around it. Wow, it's holding all the good stuff in.'" While Beverly hoped earthbound people would understand, it was clear that having the reality "put . . . in [her] face" was the trigger for her own changed understandings and appreciation of our world's beauty.

"Don," whose most significant work with NASA occurred in the 1970s, was a self-described born-again Christian whose experience of the overview effect was seen through a religious lens. When he talked about seeing the Earth from space, he described it as truly transformative. A friend of his had recently booked a space tourism flight, and Don shared the advice he'd given his friend: "Don't unstrap, just look out the window. You can do zero gravity in an airplane; pay 5,000 bucks and you can go zero gravity. But you'll never have a view like you're going to have by looking out that window to see the curvature of the Earth and the beauty of the Earth from 400,000 feet. It's five minutes! It will change your life! Look out the window!" When Don mentioned "zero gravity in an airplane," he was referring to parabolic flights, those special airplane excursions that dive at different rates to make passengers temporarily weightless. The only way to see the Earth from space, though, is to leave the planet.

For Don, like Beverly, seeing the Earth from space wasn't just about beauty but also confronting a startling reality. In Don's experience, this reality was a supernatural one. Seeing the planet from the outside allowed him to confirm scriptural quotations, proving, he believed, that the Bible contained knowledge beyond what its human authors knew and proving its divine origin. In an interview he explained,

> There are scriptures like, "God sits enthroned above the circle of the earth." Well, with my eyes, I've *seen* the circle of the Earth. . . . When that was written 800 years before Jesus, nobody had seen the circle of the Earth, but *God* had. . . . And in the book of Job, there's a verse that says when God made the Earth, he suspended it upon nothing, and that's exactly what it looks like.[40]

Other astronauts I've spoken to have had similar awe-inspiring experiences in which gaining a whole new perspective on our world gave them a type of knowledge and certainty about the planet that they hadn't possessed before. There were different levels of certainty, but most I spoke with developed a new appreciation for the vulnerability of our planet and the life growing here. "Theo" came away with the understanding that the Earth was a "little spaceship" and we're all passengers who should respect each other. "Cara" called it a "tiny little oasis" that needs to be cared for. For "Charles," the view made him realize the breathable atmosphere was so thin that of course human-made climate change could happen and that "we might cause ourselves a great deal of trouble."

Psychologists believe that experiencing awe can have a powerful impact on people, triggering religious or spiritual beliefs in some cases, causing positive personal transformations in others. That said, awe isn't an emotion of pure pleasure. It seems to combine pleasure with pain, dread, and wonder. Guilt and shame can even be involved. For example, some astronauts have said they felt like they were doing something forbidden by seeing the Earth from space. In his 2016 memoir, astronaut Mike Massimino wrote, "As I looked down, the thought that entered my head was this is something I'm not supposed to see. This is a secret. I'm not supposed to be up here."[41] Awe is not a gentle emotion. It is a feeling that happens when a bewildering reality intrudes into your life and challenges you with its unexpectedness. The mind scrambles, grasping for explanation. Interestingly, what astronauts get from the overview effect doesn't generally contradict what they already know and believe, it just makes those things seem a lot more concrete. Fuzzy understanding clicks into sharp, undeniable focus.

But encounters with space don't always clarify things—sometimes awe overwhelms, deepening confusion, confronting astronauts with the sheer strangeness of the cosmos. It can highlight, dramatically, the limits of human comprehension. As we'll see, this sense of confronting the unknowable becomes even more profound as one shifts their gaze from Earth outward into infinite space, exploring experiences that I call the ultraview effect.

3 *The Ultraview Effect*

Gazing into the Infinite

In 1932, biologist J. B. S. Haldane wrote that the universe is not only stranger than we suppose, but stranger than we *can* suppose.[1] With this statement, Haldane underscored the idea that as terrestrial creatures, human beings may not have the sensory equipment or cognitive capacity to fully understand the cosmos. Haldane's point, then, wasn't that humankind *didn't* understand the universe, but that understanding it might be beyond our ability. But that doesn't mean that no one ever tries. Because of the nature of their work, people in the space community have ample opportunity to contemplate and confront the strangeness of the universe.

These people—whether physicists, engineers, medical doctors, technicians, astronauts, authors, or others—think about outer space far more than do the members of their wider community, reasoning through problems, envisioning solutions, and contemplating large-scale events, social or cosmic, that will affect the entire planet. Astronauts, of course, experience space as a place where they are, have been, or hope to go. Because of this, the lived experiences of astronauts can be especially interesting, given that they create a kind of extraterrestrial culture, adapting human behavior to isolation, microgravity, and the stunning vistas they encounter.

Astronauts are sometimes afflicted with a condition called spaceflight-associated neuro-ocular syndrome (SANS). This condition involves fluid build-up in the head that can cause swollen optic nerves, changes to the shape of the eyeball, lasting farsightedness, and retinal damage.[2] Because of these and other potential disabling effects, some researchers have argued that it might actually be more efficient to send people already adapted to disability, including the blind or vision-impaired, into space.[3] Although vision can be compromised by space travel, it's clear that what astronauts can *see* is often especially significant, and that sight is the crucial sense behind what Frank White termed the overview effect.[4] The impact of looking at Earth from space, though, isn't predictable. Sometimes it can be anticlimactic, as it was for Alan. It can also intensify feelings of mortality, as it did for Paul and William Shatner. For some, it reinforces religious beliefs, as it did for Don. And for many astronauts, including Beverly, gazing on the Earth makes the beauty and vulnerability of our planet clearer and more real.

The Importance of Vision in Sacred Space and Outer Space

I studied pilgrimage for many years. The impact of the visual is often important to people visiting sacred places and is a huge part of why certain locations inspire awe. For many cultures, seeing a sacred place from a distance—whether it's Jerusalem, Rocamadour, Uluru, or Mount Kailash—can be a deeply spiritual experience, but it's important to recognize that this kind of gaze is not always neutral. At times, the distant view of a landscape has been tied to power—whether in the mapping of new lands during colonial expansion, or in the way some sacred sites have been reinterpreted through an outsider's perspective rather than that of the people to whom they hold deep meaning.

Religious tourism offers a striking example of how a distant gaze can shape perception. Jackie Feldman, an anthropologist who has worked as a tour guide in Jerusalem, has written about taking groups of evangelical Protestants to places where they can see the whole of the city spread out before them, which is both "a projection of sacred text onto the land" and a way to achieve a God-like view.[5] Another example is the ancient French shrine of Rocamadour, built onto the side of a cliff in rural France. Rocamadour's structures form an array of gravity-defying churches and other buildings that somehow seem to emerge naturally from the stony cliff face but also testify to the amazing architectural feats of the site's fourteenth-century builders. When I lived in Rocamadour in the late 1990s, I learned that the view was one of the main reasons people, both religious and nonreligious, came to visit. Even my own agnostic mother said, in clear awe, that she "felt something" when taking in the entire panorama from the other side of the Alzou canyon.[6] A mostly nonspiritual woman, she puzzled over her overwhelming feelings at Rocamadour and would sometimes recall the experience, wonderingly.

Astronauts in space feel something similar. Part of the overview effect is a powerful sense of understanding reality in a new way. It's astonishing but also illuminating; space travelers who have experienced it talk about understanding weather patterns with greater clarity or realizing that national borders are both artificial and rather arbitrary when viewed from above. Akin to seeing holy sites like Jerusalem or Rocamadour from the outside, seeing the Earth—itself sometimes thought of as a sacred place—gives the viewer that sense of taking in an entire reality, grasping the whole of an important location. Astronauts who have looked at our planet from orbit have described the experience as seeing something the way a deity might or finally understanding something from a divine perspective. As a scholar of pilgrimage, the overview effect made sense to me because it was so similar to the experiences of pilgrims.

One afternoon talking to Zack, however, I learned about other phenomena that were much less familiar. He had experienced an unusual type of overview effect from looking at the Moon from lunar orbit, for instance. When astronauts talked to me about looking at the Earth from space, they tended to mention the beauty they saw: lightning pulsating in clouds, shimmering waves from auroras, meteor showers in the atmosphere below them, the shapes of islands and cloud formations, or the stunning image of the Sun rising over the Earth's dark edge. The Moon has none of those things. With no giant expanses of liquid water, almost no atmosphere, and no magnetic field, the Moon seems distinctly alien and even dead. Because of the strangeness of the experience, Zack counted seeing the Moon from lunar orbit as one of the most "profound" things he'd seen during his space travel. He explained, his voice vibrating with remembered excitement:

> You can look at the Moon and see the dark circles and all that but [when] you get close to it and you see the smaller craters, and you see how sharply defined they are and . . . the very sharp shadow patterns caused by the Sun, well you're seeing it for the first time. And the Moon's very unfriendly, it's very unforgiving, very, very difficult. . . . you kind of decide in your mind whether it's hostile or not, you know that it's not something you want to step out into. . . . You feel it, you feel the isolation. Not so much that you think it's

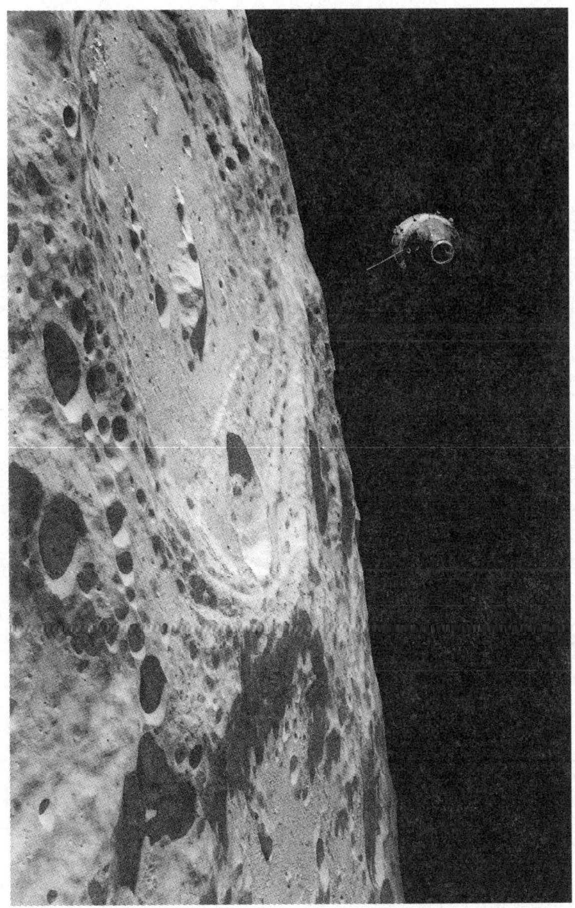

FIGURE 4. Apollo Command Module orbiting the Moon, demonstrating the vast scale and stark terrain encountered by astronauts during lunar missions. Illustration by Karl Tate.

> hostile, it's like you are so cut off from everything you know, and here, all of a sudden, you're looking at the Moon that you probably looked at thousands of times growing up, and here you are anywhere from 16 to 10 miles away from it, and it's *different*.

Zack's description of being near the Moon illustrates a unique experience of awe. The surface of the Moon—with its overwhelming stillness, rocky exterior, and stark mountains, valleys, and craters—is the ultimate desert. The Earth, in contrast, is a kaleidoscope of life, with teeming oceans, sparkling cities, and swaths of rainforest all visible to those who orbit it. Zack described it well when he said, "You see the Moon, I mean, *there*, there's nothing. And the only thing you see on the Moon is desert and craters and volcanoes, and stuff like that. But there's nothing alive. But you look back at the Earth, and it's *alive*."

The Moon, to Zack, was beautiful but barren, an object of curiosity rather than connection. This idea of viewing landscapes from a distance—whether sacred sites on Earth, the Earth itself, or the Moon—raises important questions. Who gets to have this perspective, and what does it mean to see a place from "above" rather than from within? Some scholars have noted that space exploration has been attempted by only a few powerful nations, raising the question of how different cultural backgrounds might shape the overview experience.[7]

For the astronauts I've interviewed, seeing a world from orbit fundamentally altered how they understood it. Some astronauts experienced awe as a result, others did not, but either way there was better comprehension. The lunar overview effect, like the terrestrial version, appears to make an abstract understanding come into focus. Orbiting a planet or moon (our own, or one more distant) seems to enhance perspective and clarity about what's being observed, with details that were once obscure suddenly taking shape. Seeing both the Earth and the Moon from space allows contemporary astronauts to truly know them in a way that's not possible from either a great distance or from surface level. But there are vistas in space that strip understanding away. While seeing the Moon up close gave Zack a stark sense of lifeless isolation, looking into the Milky Way was an entirely different kind of overwhelming experience.

The Milky Way Experience

A different incident gave Zack a vision distinctly different from the overview effect, in that it *defied* knowledge. This was his observation of a star field during a brief part of his lunar mission, when he was in complete shadow with both the Earth and the Moon blocking the light from the Sun. He was also, in those moments, outside the range of communication with NASA and temporarily alone in the command module. In true solitude, Zack turned off all the lights in his spacecraft, allowed his eyes to get used to the darkness (becoming "dark-adapted"), then looked out the window into the Milky Way galaxy.

The Milky Way is our own galaxy, a spiraling conglomeration of stars, planets, meteors, comets, black holes, and more. If you can imagine a creature like an octopus spinning so that its tentacles spiral around it in the same direction, you'll have some idea of the shape. Our solar system is located in one of these tentacle-resembling sections known as the Orion-Cygnus Arm, about halfway between the galaxy's center and the outside edge of the circular shape created by the arms.[8] In our location we are surrounded by a vast array of stars, and while we can see some of them from Earth, there are many, many more we cannot see (although using technology to "see" infrared, radio, and X-rays has helped us find some of these).[9] While Zack could not peer into our galaxy with the accuracy of the James Webb Space Telescope, letting his eyes adjust to the dark and looking out at the stars with his vision unhampered by Earth's atmosphere gave him a view he felt was substantially different from seeing the Milky Way from Earth.

His profound moment of quiet solitude, gazing into the depths of space, was a topic we returned to multiple times. It seemed to have established a stronghold in his memory, and his eyes looked distant, like he was seeing it again, when he said, "I was in complete darkness, at that point, the stars pretty much filled the sky. Because I'm looking at them, without any interference. Without any atmosphere or anything else in the way. So you see them all." One of the things that looking into the Milky Way forced Zack to contend with were the concepts of time and space without end. Zack's experience—staring into an overwhelming field of stars, unable to grasp their sheer vastness—perfectly fits what Edmund Burke described as the sublime. True awe isn't just about beauty; it comes from confronting something so immense that it's unsettling.[10]

Take a moment to imagine a solitary astronaut, in darkness and, for a few minutes, out of communication with NASA, looking out into a field with so many stars that his brain cannot process them. He is deeply, overwhelmingly alone, and his spacecraft window is filled with light. "It changed my ideas of infinity," Zack said. "It changed my ideas of time. There's no beginning of time. There's no such thing. . . . Well, infinity is just something beyond what we can contemplate. So that changes your outlook on everything."

Zack was blown away by the experience, thrust into doubt about what he thought he knew. Burke's description of a mind filled with "delightful horror" seems like it fits Zack's experience very well. He returned to Earth a few days later, but what he had seen interfered with his sleep, compelling him to research what he didn't understand and record his impressions.

Zack's encounter with the stars was rare—an experience difficult to replicate and fundamentally different, I realized, from the most common version of the overview effect. An astronaut orbiting Earth (or the Moon) can see geographical features very clearly. Low Earth orbit provides a display of weather patterns, islands, continents, vegetation, and evidence of humankind. Astronauts look out the window (the beautiful cupola on the ISS was built for this purpose) and increase their understanding. Looking out into space may have also increased Zack's understanding to some degree. Certainly his mind was forced to reckon with the enormity and density of the star field he saw. Despite this, the main thing the vision of space did was push Zack to realize how much there was to know and how very little he *did* know. This was not the overview effect. Zack was looking beyond the Earth, beyond the solar system, beyond even the Orion-Cygnus Arm of our galaxy. Latin for "beyond" is *ultra*. Inspired by Frank White's overview effect, I began referring to what Zack had described as the ultraview effect. While the overview effect refers specifically to astronauts seeing Earth from space and experiencing clarity about our planet's fragility and interconnectedness, the ultraview effect describes the overwhelming sensation astronauts feel when gazing out into the Milky Way, confronting humanity's profound ignorance and the immense unknowns of the universe. In this way, the ultraview effect pushes awe beyond clarity into a more unsettling realm of vast mystery.

Zack was changed by what he saw. When he returned from his mission, he found himself consumed by a lot of social activities, but instead of enjoying them he chose to break away. Late at night he would regain his solitude and think about what he had experienced in lunar orbit. He said, "I'd sit in my living room, and all these thoughts would come pouring through. So I began writing them down. . . . Could never understand it. It was like I was being guided by something. I don't know. It's kind of an open channel or something like that." Although he suspected most of that feeling had to do with his deep levels of physical and mental exhaustion, he really did feel powerfully transformed and that his experience in space "broke down the barriers in my mind."

Seeing the galaxy as a "sheet of light" denser and brighter and more awe-inspiring than he'd been able to see it from Earth seems to have transformed Zack in a fundamental way. That transformation wouldn't come as a surprise to Theo, a shuttle-era space voyager who also experienced the ultraview effect.

Theo, like Zack, was raised in a mainline Protestant church. Although his family was not particularly active in their religious pursuits, he felt from a young age that spirituality was one of the core parts of his personality; he was always interested in a reality beyond what he could see. His eyes bright, giving a feeling that he

---

FIGURE 5. A detailed comparison of the overview effect and the ultraview effect. While the overview effect emphasizes viewing Earth's unity and vulnerability from space, fostering environmental consciousness and responsibility, the ultraview effect emphasizes observing the vastness of the cosmos, promoting humility, philosophical reflection, and a heightened awareness of human ignorance. Illustration and table by Karl Tate.

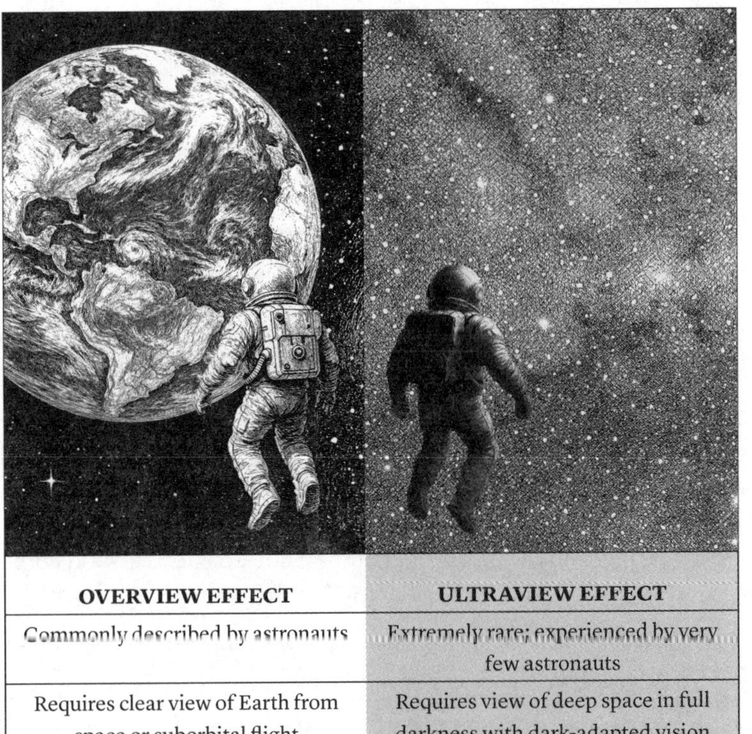

OVERVIEW EFFECT	ULTRAVIEW EFFECT
Commonly described by astronauts	Extremely rare; experienced by very few astronauts
Requires clear view of Earth from space or suborbital flight	Requires view of deep space in full darkness with dark-adapted vision
Triggered by viewing Earth from space	Triggered by viewing the Milky Way galaxy from space
Visual emphasis on Earth's beauty, unity, and fragility	Visual emphasis on vast star fields and the infinite cosmos
Emotional experience of clarity, connectedness, and protectiveness	Emotional experience of awe, humility, and profound uncertainty
Reinforces awareness of Earth's vulnerability and interconnectedness	Highlights limits of human knowledge and the universe's vast scale
Leads to increased environmental consciousness, responsibility, and unity	Leads to philosophical reflection, deep humility, intense curiosity, unsettling awe, and a heightened awareness of profound human ignorance

was still young no matter his biological age, he told me about his boyhood nature-based spirituality:

> The creatures were sacred, the stars, the world. . . . So this world I'm out in, this is a sacred world. It was sacred. The trees are sacred, everything was sacred. And a three-year-old in the middle of a forest at night is a real adventure. But I had the faith, I had the faith that the spirits would look after me, not only not *harm* me, but would look after me. So I guess as a child, I was both a pantheist and [a Protestant].

Theo's religious leanings were eventually expressed in a nearby community church. When he was about twelve, he became an acolyte working with the minister, taking on new religious responsibilities. He had kept some of his more unorthodox ideas, however, and told me that as a child he would open the window in his house, kneel reverently, and pray, eyes open, while looking at the starry sky.

Theo has had a curious mind his entire life, and as he grew older, he began to learn more about other religious traditions. When I interviewed him, he described his spiritual practices as a blending of Christianity, pantheism, and Buddhism. Perhaps unsurprisingly, Theo's boundless curiosity inspired him to study multiple academic subjects, learn a diverse set of physical skills and talents (like piloting planes), and become absorbed in a wide range of approaches to engaging with the spiritual world. Of the astronauts I have interviewed, he is the only one who seemed specifically set on participating in rituals and religious practices while in space, including gazing, dark-adapted, at the Milky Way from a spacecraft.

Theo's search for spiritual meaning in the heavens is probably rooted in his lifelong interest in the stars. In addition to observing the starry sky when he prayed as a child, Theo also valued the stars in navigation, explaining to me that in the early days of crewed spaceflight, an astronaut's knowledge of the locations and names of individual stars and constellations could be used to recover the computer's navigation system in the event of a problem. Certain star formations meant a lot to him as well, with the Southern Cross moving him spiritually. He reflected on whether he had subconsciously linked the Southern Cross to the Christian cross.

Unsurprisingly then, give his passion for stars, Theo had a strong desire to see them with his own eyes while in space. Theo only experienced the ultraview effect, however, after convincing the rest of the crew on one of his later missions to let him do it. I could hear the frustration and disbelief in his voice when he described a failed attempt to persuade an earlier crew to take the necessary steps, which were to turn off all the lights in the spacecraft and look out the window. He said indignantly, "I was never permitted to have a dark experience because those people were not interested, and they wouldn't let me turn all the lights off. They don't care. They don't want to look out the window. You can see *galaxies* with the naked eye, galaxies!" Many years later, Theo still couldn't believe it.

On a later flight full of "incredibly considerate people," he finally got his wish during two "night passes" that took place when the spacecraft was on the dark, unlit side of the Earth where it was night. Theo's telling makes it very clear why the ultraview effect is so rare. He explained that normally during a night pass, no one can look out the window into space because the spacecraft, like most other workplaces, is full of light.

Some evening, try looking out the window in a brightly lit room to see the night sky. You'll find it's easier to see reflections of what's in the room than any stars outside. What's more, in a lit room a person's pupils constrict to match the level of the interior lighting, making dimmer objects like stars fade into nothingness. Orbiting astronauts who look down at the Earth, especially in daylight, will see the planet very clearly as the continents below the space station are bathed in brightness. Looking away from the Earth into the Milky Way, however, is much more of a challenge.

Theo joked that even in space, when you can't clearly see anything through the window, your impulse is to just "go outside." Of course astronauts can't just hop outside when they want to get a better view. The process of getting ready for a space walk, which can't be done impulsively, takes between two and three hours.[11] This meant that full crew cooperation was absolutely necessary for Theo to look out the spacecraft windows. It's no wonder the ultraview effect is so uncommon.

Fortunately for Theo, his chance to look outside finally happened. As he expected, when he saw the Milky Way through the spacecraft window he was blown away. "It's a hard wall," he told me, "It's a hard wall. The Milky Way is a hard white wall. It's unbelievably solid up there." He and one of his friends on the crew celebrated by taking communion as they floated together in a pitch dark spacecraft gently illuminated by the light of a "hard white wall" made up of billions of stars. What Theo experienced was rare, but another astronaut I talked with got a small taste of it during a different mission.

I first met with "Charles" on a spring afternoon. As we got acquainted, we talked about how astronauts had their own "jocks versus nerds" elements just like high school—Charles, gentle and professorial, was a self-identified "nerd." A shuttle-era astronaut, Charles had once seized a rare opportunity—an unplanned moment of downtime—to gaze out at the stars during Earth orbits.

His approach to achieving this extraordinary view was both creative and impulsive, taking advantage of the chance to let his eyes adapt to the darkness. He described how one of the shuttle's windows had Velcro around its frame, designed to attach a special camera bag that shielded the lens from ambient spacecraft lighting, thus allowing clearer photos of space. Charles creatively repurposed this setup by climbing partially into the camera bag himself and fastening it around his waist. "I became the camera," he explained, meaning that he used the bag to block out surrounding light, immersing himself in the darkness and clarity of space.

While floating in microgravity with his upper torso secured in the camera bag, Charles adjusted to the darkness. He told me, "After you dark-adapt enough then you get to see the colors of the stars in ways that I couldn't have seen before." He spent two orbits in the camera bag, or roughly three hours, and what he saw during that time varied:

> Much of it was looking at the Earth, but . . . also looking out at the stars. . . . You're much more aware of how many there are. It's just everywhere. It's just dense. But it's also colorful. I can see Mars's red and can see Betelgeuse's red and some of these things. You don't see much . . . color from the Earth. But up there it was much more obvious. And we were at a high enough inclination that I could see the aurora over Antarctica. . . . And you could see the waves in the aurora. It was in early August, so while I'm watching I can see shooting stars below me.

It's very clear that for Charles, looking out into the "dense" array of colorful stars was impressive, but not more so than the Aurora Australis (also known as the Southern Lights) or the Perseid meteors passing through Earth's atmosphere. Charles's story reveals, however, that the greatest contribution to having awe-inspiring visions in space is probably dark adaptation.

Without dark adaptation, looking out into space from a spacecraft isn't possible with the naked eye. Neither Zack's star field, Theo's hard white wall, or Charles's visions of the star Betelgeuse and the planet Mars would have been as detailed as they were without it. All three men allowed their pupils to expand, letting in as much light as possible, creating a situation in which even very dim light takes on added heft and dimension. While discussing what he saw while dark-adapted, Charles said,

> There was an awe about it. . . . You can *talk* about how many jillions of galaxies there are and how big each galaxy is. But I think the view from there gives you a little bit more sense that this is a big place and there's a lot out there. People have said, and I guess I would share this, that there's a sadness about being in our time because you won't know what the next time knows.

Like Zack, the vision of stars from space made Charles deeply aware of what he did not understand and envious of future humans who would probably have a much better comprehension of the cosmos.

The view Zack, Theo, and Charles could see from space was entirely different from what they could see from Earth. This was why, Theo said, telescopes like the Hubble are put out into space where they don't have to contend with our planet's atmosphere. In another interview, I asked Theo his conclusions about the impact of seeing the Milky Way while dark-adapted in space. He said the point of space exploration should be for humanity, not test objectives, and that being in the stars enriches spirituality like nothing else—if space doesn't create spirituality, what will? He continued, the passion rising in his voice, "You're looking at galaxies with the naked eye. You're in the heavens. . . . You're in a place that will increase spirituality. You'll do the checklist a hell of a lot better if you're on top of your spirituality than if you're not. It's the same as . . . water or food." If spiritual satisfaction makes better workers, then space, at least according to Theo, is a shortcut to that satisfaction. Theo was arguing that quality of life, important in all workplaces, could be heightened for crews through the spiritual and religious opportunities available in space. If NASA were receptive to this kind of thing, he suggested, they'd have happier, more efficient astronauts.

Implications of the Ultraview Effect

What does the ultraview effect do to people? So few have experienced it that it's impossible to know for sure. Any chance to experience the ultraview effect is extremely limited—astronauts are typically kept so busy with lists of assignments that having the downtime (and cooperation from the rest of the crew) to stop, get dark-adapted, and look out into space is very difficult. At least the ISS has the cupola, a module of the space station with large windows that encourages astronauts to gaze at Earth. But looking at Earth doesn't require darkness or dark adaptation. All of this explains why the only people I've interviewed who had looked into space while dark-adapted were men—something that may not just reflect chance but also systemic barriers. Women in space have historically faced higher expectations to prove their competence, which may have left them with fewer opportunities for unscheduled reflection or dark adaptation.

Shuttle-era astronaut "Cara," for instance, had been concerned that any failures on her part would reflect badly on other women, explaining, “I thought if I made a mistake . . . Mission Control and maybe even the national media would say, oh, that’s what happens when you have a woman [astronaut]!” As a result, she told me, her mindset during her time in space was, “I’m not going make a mistake. I am focused like a laser. I’m doing this job perfectly because of that.” This impacted her experience. She explained, “I didn’t really enjoy my flights as much as I think I could have. I mean, if I went along as a passenger, I would’ve had a really great time, but I was working and I was focused and I was constantly thinking what’s next, what’s next on the flight plan.” Under these circumstances, taking time to gaze at the Milky Way would have been completely unworkable.

For multiple reasons, then, experiencing the ultraview effect is rare, but my preliminary research can be summed up by something Zack said. He realized, after encountering the immensity of the star field he saw during his mission, “We don’t know crap about anything. We really don’t.” By confronting astronauts with the apparent reality of infinity and forcing them to reckon with a galaxy that dwarfs our planet and our population, the ultraview effect may act as something like a cosmic reset, encouraging astronauts to return to a more childlike state of understanding in which their sense of knowing exactly what they are doing and feeling confident about their reality is somehow shut off.

Awe dismantles our feelings of certainty. Zack’s encounter with the ultraview effect echoes how many astronauts describe the overview effect—not simply marveling at beauty but confronting the unsettling yet invigorating realization that there is far more to reality than previously understood. While awe can spark curiosity, curiosity alone isn’t sufficient for true exploration. Before the rush to fill gaps in human knowledge, there is an essential intermediate step: humility. Awe helps trigger humility, reminding a person of their smallness and gently—or sometimes forcefully—bringing one face-to-face with their ignorance. Although setting out on a journey confident that one knows the answers might feel reassuring, the greatest discoveries come from an openness born of humility. In the chapters that follow, we'll take a deeper look at humility itself: what it means, why it's valued in cultures across the world, and how it benefits genuine exploration.

II *Humility*

4 *The Nature of Humility*

Examining a Virtue

In early March 2019, I flew to California to do ethnographic research with robotic spacecraft engineers and programmers at NASA's Jet Propulsion Laboratory (JPL) in Pasadena. JPL is "a leader in robotic space exploration, sending rovers to Mars, probes into the farthest reaches of the solar system, and satellites to advance understanding of our home planet."[1] The campus of JPL resembles a large university, with multiple buildings, each looking like it could house an English department or a chemistry lab. Some hold mission control centers for supervising distant space probes; others serve as points to translate and analyze the messages the spacecraft send to Earth; and still others are like science fiction workshops or factories, where white-garbed technicians in highly sanitized clean rooms meticulously assemble Martian rovers and other wonders of the space age.

One of my favorite spots was the Mars Yard. A friendly and soft-spoken JPL engineer I will refer to as "Alonzo" gave me a tour of the campus on a clear late winter day. We made our way up a hill to visit the testing area, where a replica is kept of the Curiosity rover NASA sent to Mars in 2012. Signs around the Mars Yard indicate the replica is called MAGGIE, which stands for Mars Automated Giant

Gizmo for Integrated Engineering. Curiosity's handlers deal with what they don't know by putting MAGGIE through different conditions, like navigating rocky paths or slippery sand, or completing certain tasks here on Earth before its Martian counterpart, Curiosity, attempts them on the red planet. JPL was a gorgeous place to visit, and I was awed by its dedication to increasing the still limited human understanding of our solar system and beyond.

During that same visit I headed to La Cañada Flintridge, a small city northwest of JPL, where many of its employees have homes. I had a meeting there, in a cozy Southern restaurant, to interview a spacecraft computer expert I will call "Cody" (and yes, that's a pun on computer coding). Cody's combination of rapid intellect and relaxed demeanor made him especially fascinating. Although many of the brilliant engineers and scientists I've interviewed demonstrated intellectual humility, Cody stood out for the remarkable clarity and depth with which he articulated his acceptance of uncertainty. Over our three-hour conversation, he was notably comfortable acknowledging what he didn't know. For Cody, openly admitting these gaps in knowledge wasn't merely honest—it was essential to his moral worldview.

Cody was raised in the Midwest in a conservative Christian religious tradition, one that didn't want children asking questions or raising doubts about what they were taught in church. As he finished high school and went on to college, Cody became more and more dedicated to creating an understanding of the world (and universe) using factual information he could verify himself rather than relying on the assurances of religious leaders, supervisors, or other authority figures. If factual information wasn't available, or wasn't available *yet*, he took that seriously, unafraid to admit the missing facts and sit comfortably with the knowledge that there were gaps in his understanding. For Cody, pretending to know something he didn't or taking something on faith that he couldn't prove would have violated his sense of morality.

As we discussed religion, spirituality, and science, Cody told me he sometimes imagined that the universe involved a "creative force" beyond his understanding. He explained:

> I mean we don't have another word to describe it . . . you could call it *the universe* but somehow that doesn't seem like it's a big enough word for it. So, you could then venture out into religious-type terminology. Some sort of pantheism, you know, like God is everything and everywhere. And I don't really have a problem with people wanting to think of it that way because your definition of God is just so *vague*. It kind of *has* to be because the universe is like 99.9% unknown to us. So, it's a word to describe the unknown basically.

Cody's admission that humans know so little was particularly striking. After all, his job required knowing Mars intimately enough to ensure that a rover sent there would operate exactly as intended. Cody, however, put the exploration of Mars into perspective. He acknowledged that sending robots there is extremely challenging, yet emphasized that Mars itself is both very close and very small compared to the enormity of the reality beyond it. To illustrate the sheer vastness of the universe compared to Mars, Cody offered a metaphor that scaled cosmic distances down to something familiar—planet Earth:

> It's just a big deal to explore Mars. It's so hard and it's so strange and yet it's like we haven't even walked out our front door yet. Mars is like the pebble that's on the little lip of the door frame. And the universe goes all the way out across Southern California and over

FIGURE 6. A visual metaphor of Mars as the pebble on Earth's doorstep, highlighting humanity's first small steps toward space exploration. Illustration by Karl Tate.

the ocean. So the vastness of it is just so amazing and just terrifying with the powers involved, and the sizes of things involved.

Contemplating the vastness of the universe clearly evoked awe in Cody, reinforcing his awareness of how little he truly knew. As we've seen, Edmund Burke's notion of the sublime similarly blends fear and wonder. Burke, like Cody, believed encountering the sublime could leave us feeling profoundly small. In his work *A Philosophical Enquiry into the Origin of Our Ideas of the Sublime and Beautiful*, Burke wrote about the human contemplation of God (perhaps similar to Cody's notion of a creative force): "But whilst we contemplate so vast an object, under the arm, as it were, of almighty power, and invested upon every side with omnipresence, we shrink into the minuteness of our own nature, and are, in a manner, annihilated before him."[2]

This idea that our awe-filled reaction to the sublime can make us "shrink into . . . minuteness" is what the beginnings of humility feel like. A sense of being so small that you basically don't exist is not pleasant. Most humans want to feel significant. Cody, though, who spent both work and leisure time contemplating the unknowable universe, embraced that feeling. As we ate our sandwiches in the shadow of JPL, he noted that his willingness to choose a feeling of insignificance over accepting the unprovable stories of his childhood put him at odds with some friends and family, stating, "That's the thing that I've always stuck to that people get frustrated with. I think it's *okay* not to know things. I would much rather just admit to myself that I don't know something than to try to come up with an answer or convince myself this answer is correct. I think it's better just to *not know*." Cody's willingness to be humble is a good place to start our examination of the meaning of humility.

Defining Humility

Humility can be a complex concept, varying in meaning across cultures. At its core, humility involves having a modest or realistic view of oneself, free from arrogance or excessive pride. Importantly, this is distinct from humiliation, which involves embarrassment and shame typically caused by external actions. Being humble suggests an internal sense of being "grounded," comfortably aware of one's own limitations without negativity or embarrassment. A person can thus be both confident and humble—a compelling combination.

Concepts similar to the idea of humility can be found across cultures. In Tagalog, one of the languages of the Philippines, the word *pagpapakumbaba* can be translated as humility; its root word, *baba*, means "low." It is considered one of the primary virtues in Philippine culture, particularly in religion and in business, where an entrepreneur with the opportunity to succeed "no matter how lowly the task, [will be found] at work patiently, and unmindful of his social status."[3] Likewise, in Samoa, *loto maualalo* (humility) is seen as a core virtue alongside *alofa* (love) and *fa'aaloalo* (respect), with all three connected to the notion that everyone has "a divine designation or role to play in society that can benefit and enhance the community they live in."[4]

Humility is also a central concept in much of the Americas, although it may differ slightly from its roles in the Philippines and Samoa. For instance, a group of public health researchers in Canada conducted a study on health and wellness with different Indigenous North American communities. The researchers took great care in their investigations to incorporate what they identified as the "Seven Grandfather Teachings" of eastern North

American peoples like the Lenape, Algonquin, Ojibwe, Odawa, and Delaware. These teachings were specifically "respect, honesty, truth, humility, courage, wisdom, and love," with humility defined as having "compassion and care for all of creation and understand(ing) that all life is equally important."[5]

Are There Benefits to Humility?

It's clear that many communities worldwide intentionally cultivate humility in their members. Why is humility considered so beneficial? Anthropological research suggests that humility is highly valued because it strengthens social connections, enhances cooperation, and even helps groups solve problems. Anthropologist Michael Lambek observed that among the *fundi*—village experts on the French island of Mayotte, off the coast of East Africa—leaders are expected to be "patient, courteous, and modest. . . . [They] should not be conceited or make too many conditions, nor put on airs."[6] Arrogance is off-putting because it raises suspicion about whether a leader genuinely knows as much as he claims. Similarly, Richard Lee found among the Dobe Ju/'hoansi people of southern Africa that skillful hunters are expected to demonstrate modesty. Although some hunters clearly outperform others, their behavior is intentionally shaped by the community "to minimize the tendency toward self-praise and to channel their energies into socially beneficial activities."[7]

But how exactly does humility benefit human communities? Sociobiologist Robert Trivers offers one perspective with his theory of reciprocal altruism, the idea that individuals help one another because they anticipate receiving help in return, directly or indirectly. Trivers argued that humble, altruistic behaviors build networks of mutual aid, allowing communities to thrive through cooperation rather than competition.[8] If you've ever had a friend bring you soup when you were sick or help you move, you've experienced the benefits of reciprocal altruism. Not everyone agrees that reciprocal altruism fully explains why humans value humility and cooperation, however. Scholars like E. O. Wilson and David Sloan Wilson (no relation), for example, emphasize group selection, arguing that cooperative behaviors are shaped because they benefit entire communities, not just individual relationships.[9] Anthropologist Joseph Henrich argues that cultural norms may be even more influential than biological instincts in guiding human cooperation.[10] Despite these debates, Trivers's foundational insight remains compelling: humility and cooperation are powerful social assets.

Do these benefits of humility extend to romantic relationships? Evidence suggests they do. A study led by Ruoxi Chen found that both American and Chinese adults valued a personality trait known as "honesty-humility" in potential romantic partners, although other preferred traits varied culturally.[11] This emphasis aligns with the HEXACO model of personality developed by Kibeom Lee and Michael Ashton, which identifies honesty-humility as crucial for trustworthiness[12]—indicating that humility is valued even in deeply personal relationships.

Still, there's an important distinction between the humility valued in social interactions or romantic relationships and the intellectual humility exemplified by Cody and Zack. The humility discussed thus far involves modesty, cooperative behavior, and trustworthiness. Intellectual humility, however—the central focus of this book—is specifically about acknowledging the limitations of knowledge and being comfortable with uncertainty and the unknown.

Intellectual Humility

Psychologist Elizabeth J. Krumrei-Mancuso and her colleagues have studied intellectual humility extensively, connecting it directly to the process of learning. They argue that humility "in general, is often thought of as a virtue relevant to gaining new knowledge, as most definitions of humility involve acknowledgement of one's limitations. Simply put, learning requires the humility to realize one has something to learn." They define "intellectual humility" specifically as humility "in the domain of thoughts, beliefs, ideas, and opinions," characterized by "a non-threatening awareness of one's intellectual fallibility."[13] This precise form of humility is exactly what we observed earlier in astronaut Zack, awestruck by his vision of the star field from space, and in Cody, whose contemplation of the universe from Earth similarly revealed to him how little humanity truly knows. Both experiences highlight the essence of intellectual humility—a clear-eyed awareness of our own ignorance about much of reality.

Yet, I suspect that openly espousing intellectual humility might make some readers uncomfortable. Isn't it preferable, many might ask, to appear smart, confident, self-assured, and wise? Admitting self-doubt or openly acknowledging a lack of knowledge can feel vulnerable—almost like admitting weakness. As a college professor, though, I've had the experience of reading an answer to an exam question or even an essay where a student clearly doesn't know the right answers but continues to forge ahead, creating a response that might *sound* right in the hopes that I won't know any better (given that I write my own tests, this isn't a wise bet). I've also seen politicians answer questions on unfamiliar topics by bluffing their way through them, hoping to impress journalists or at least potential voters.

This pretense of certainty might occasionally inspire temporary trust, but it carries serious drawbacks. Most notably, it severely limits genuine learning—if you pretend to know something, you lose the chance to actually discover it. Further, a lack of intellectual humility isolates you from valuable perspectives, creating interpersonal conflict and hindering collaboration. Thus, despite the discomfort involved, the advantages of intellectual humility outweigh the risks of embarrassment or appearing weak.

Political scientist Philip E. Tetlock's research into the accuracy of expert political predictions vividly illustrates this point. Tetlock distinguished between two types of forecasters: "foxes," who remain open-minded and draw on multiple sources and viewpoints, and "hedgehogs," who rely on a single theory or framework. While hedgehogs can appear confident and persuasive, foxes—those with intellectual humility and openness to multiple viewpoints—consistently produce more accurate predictions.[14] Tetlock's research underscores the dangers of confirmation bias, defined as the human "tendency to process information by looking for, or interpreting, information that is consistent with their existing beliefs," a blind spot that frequently prevents deeper, more accurate understandings.[15]

Further bolstering the value of intellectual humility, psychologist Tenelle Porter and her colleagues found that individuals scoring high in this trait exhibited numerous positive social outcomes. Specifically, intellectual humility was strongly associated with openness to rival ideas—even those arising from different religious or political perspectives—as well as greater willingness to cooperate, enhanced forgiveness, and other valuable social skills. Porter's team concluded: "Intellectual humility is positively associated with multiple values, including empathy, gratitude, altruism, benevolence, and universalism, which suggests that people with greater intellectual humility are more likely to value and care about the wellbeing of others."[16] Conversely, those who overestimate their knowledge risk misunderstanding reality and creating unnecessary conflict. Given these findings, intellectual humility emerges not merely as an admirable personal quality but also as a critical social skill, beneficial for both personal insight and interpersonal harmony.

People involved in space exploration—NASA astronauts, Vatican Observatory astronomers, private-sector test pilots—are no exception to these benefits. Although not every individual in the space community fully embodies intellectual humility (indeed, arrogance can appear anywhere, even among high-profile space entrepreneurs), it remains highly valued in selection processes. Beverly, who once selected astronaut candidates, explained her personal criterion clearly: "When I interviewed people to be astronauts, my whole thing was would I want to spend six months in a small place with this person? . . . You could be the smartest person resume-wise, but if I wouldn't trust you to have my back and for you to know I'd have yours, then I don't want to fly in space with you."

Beverly's perspective directly aligns with the social qualities Porter's research associates with intellectual humility, emphasizing cooperation, openness, and empathy. These traits are particularly beneficial in environments where recognizing the limits of one's own understanding can mean the difference between mission success and failure—such as space exploration. Yet, intellectual humility isn't limited to astronauts. The entire space community, from engineers to astronomers, regularly encounters situations that challenge its understanding of reality and require its members to confront just how much remains unknown.

Humility in the Space Community

For most of us, the fact that we are on a planet, spinning in space while orbiting a star that is also in orbit around the center of the Milky Way galaxy, which itself moves at incredible speeds through an ever-expanding universe, isn't something we spend much time considering. Many of us just don't think about it—at least not most of the time.

Sometimes, however, experiences remind us of this cosmic reality. Although the film *Gravity* received mixed reviews from the space community, I vividly remember my appreciation for it when I first saw it in an IMAX theater, wearing 3D glasses. Despite some adult themes, I was eager for my son Luke to experience it as well. A few days later, I returned to the theater with my then-five-year-old in tow. With our futuristic eyewear in place, we were transported into a remarkably realistic space walk. The immense screen and three-dimensional imagery gave us the sensation of floating alongside the astronauts. Although the opening scene was intense for Luke—prompting him to hold my hand throughout the film—I cherish this memory of us floating through space together.

Yet neither watching *Gravity* nor riding Space Mountain truly deepened my awareness of the universe's enormity; it was a day spent at the Kennedy Space Center Visitor Complex, near Titusville, Florida, that truly made the vastness of space feel concrete and undeniable to me. Exploring the grounds, viewing rockets, hearing stories about humanity's journey to the Moon, and standing before the incredible Space Shuttle Atlantis made the American space program—and outer space itself—feel intensely real. Documentaries depicting the immensity of the universe and potential dangers from cosmic events reinforced this powerful awareness.

As an anthropologist who studies tourism and pilgrimage, I was fascinated by how Kennedy Space Center conveyed space exploration through its exhibits. When I studied the French shrine town of Rocamadour, I examined how priests and tourism workers influenced visitors' perceptions. Similarly, at Kennedy Space Center, the curated stories about space deeply shaped my experience. In Florida, visitors might engage with Disney, beaches, or space, among other options. At Kennedy, the clear focus was on what space truly is and how humanity interacts with it. This experience made me wonder how much more profoundly space professionals—astronauts, scientists, and engineers—might contemplate the vastness of space, and how this contemplation shapes their understanding of reality.

Cognitive Explorations of Humility in Space Science

An astronaut's experience in space isn't merely vicarious or theoretical, like watching a movie or enjoying a day at a theme park—it's real and profound. While astronauts vary widely in personality, some embody humility, although they frame it in different ways. Many astronauts downplay their achievements, attributing success to luck, teamwork, or even divine intervention.

One notable example of religiously framed humility is "Tom," a NASA astronaut with more than three decades of experience. Tom, who described himself as evangelical, dedicated considerable time after retirement to speaking with religious communities about how his space experiences reinforced the compatibility of faith and science. He viewed his entire career as guided by divine intervention, attributing little of his success to personal merit. Recalling academic struggles that eventually led him into engineering, Tom told me, "I said gee whiz, you know, this is horrible. I'm not going to be able to do this! I've got to get into something I can figure out. Maybe I'll go into engineering. But I think this was God talking to me." Even his NASA selection, Tom believed, was orchestrated by God, not because of his own abilities.

Tom's humility intensified during spaceflight. He described how even record-setting distances—such as the Apollo astronauts' journeys to the Moon—seemed trivial from God's perspective:

> Just pretend for a moment you're God, and . . . so here's these three astronauts. They're setting a record. And they're up there for two months, going 400,000 miles every day, and so at the end of two months they've gone 25 million miles. In terms of light years, that's like just a couple of seconds. And so if you were God looking at this whole universe, three guys just set a record for distance, in his eyes they haven't even gone anywhere. And if you think of time from eternity to eternity like God does . . . the guys up here for two months, what's that with respect to eternity?

Despite Earth's cosmic insignificance ("off on the edge going around a mediocre star"), Tom believed humans have special meaning to God. He concluded, "We can be as close to God anywhere [on Earth] as we can out there, anywhere in the universe. And we should feel that that is so. Because God thinks in a lot bigger terms than we do."

Cara, another astronaut, displayed humility through a similarly religious lens, but framed by her Catholic faith. During her shuttle-era missions, Cara carried cross necklaces as personal religious tokens. Reflecting on astronaut Jim Lovell's famous gesture of blotting out Earth with his thumb during the Apollo missions, Cara exclaimed:

> If that doesn't have an impact on you, you're just not thinking right! . . . I mean, we are minuscule individual people, little dots on the outside of this sphere that is rotating. . . . And then this whole thing is like going around the Sun and then the Sun is like this one tiny little thing in the galaxy. . . . But not everybody thinks about it right this way. If you actually go out there, like *holy cow*, there's a lot more to this world than just me!

Cara argued emphatically that recognizing humanity's smallness within the cosmos was essential. Her perspective, explicitly framed through her Catholic beliefs, connects directly to common human psychological responses to the vastness of space.

Both Tom and Cara use their faith to make sense of intense or overwhelming experiences. Their religious perspectives help them handle and find meaning in situations that could otherwise feel frightening or unsettling. Even if you don't share their specific beliefs, it's easy to understand how spirituality can offer comfort and resilience when facing the vast and mysterious universe. Understanding the transformative power of experiences like those of Tom and Cara raises important questions about how profound insights arise. Educational theorist David A. Kolb's concept of experiential learning offers a useful lens, emphasizing how direct, personal experiences lead to deep understanding.[17] When astronauts look at Earth from space or gaze into the cosmos, their direct experiences lead to profound emotional shifts.

However, fewer than a thousand people have actually been to space. For those of us who can't experience space firsthand, are there other ways to feel humility and awe? Cognitive psychologists have studied alternative ways of gaining insight and emotional understanding without direct sensory experiences. Among the most influential efforts in this area is Bloom's Taxonomy, developed in the mid-twentieth century by educational psychologist Benjamin Bloom and his collaborators.[18] While not without debate or criticism in the decades since its introduction, Bloom's Taxonomy has become a widely used framework to describe cognitive learning, emphasizing processes such as remembering, understanding, applying knowledge, analyzing, evaluating, and creating or synthesizing new ideas from existing ones. Both experiential and cognitive forms of knowledge have considerable value, but astronauts help illustrate a crucial difference: experiential understanding of the Earth and outer space often strikes people suddenly and vividly, while cognitive understanding generally develops slowly and deliberately, through accumulating factual information, analyzing it, and integrating it with other knowledge.[19]

Nevertheless, cognitive understanding can ultimately lead to deep insights comparable to those sparked by experiential learning. Some scientists in the space community I've met have reached a similar level of humility as astronauts like Cara and Tom through sustained intellectual engagement. Many of these individuals navigate the intersections of personal belief and professional expertise rather than adjudicating truth claims. In fact, far more scientists than stereotypes suggest find meaning in their work through religious lenses. Observing this interplay offers valuable insight into how humans, even those trained in empirical reasoning, confront the uncertainties of the cosmos.

For instance, the Vatican, through its Vatican Observatory, supports scientific research in astronomy and physics, motivated by the Catholic Church's long-standing interest in understanding the universe as part of God's creation.[20] The Vatican Observatory employs trained scientists—often Catholic priests who are also professional astronomers and physicists—to conduct research and engage with contemporary scientific debates. One such Vatican physicist, a Catholic priest whom I refer to as "Giorgio," is an expert on the space-time continuum. Giorgio spent nearly every working day contemplating weighty scientific questions such as the origin of the universe, the motion of galaxies, and the properties of black holes. Through this intensive intellectual engagement, Giorgio was keenly aware of the limits of human knowledge and his own understanding.

I met Giorgio at the Vatican Observatory, located at Castel Gandolfo. While physically outside of Rome, Castel Gandolfo is considered part of Vatican City, extending the Vatican's territorial jurisdiction beyond the commonly recognized area around St. Peter's Basilica. Giorgio exemplifies how rigorous cognitive reflection can inspire profound humility. Giorgio, like his colleagues at the observatory, holds advanced degrees in physics, publishes peer-reviewed research, collaborates internationally with other respected scientists, and attends the same academic conferences and engages with the same scientific questions as his colleagues at institutions such as NASA or the European Space Agency. The Vatican Observatory, where he works, is an established research institution that contributes significantly to mainstream scientific debates about cosmology, astronomy, and fundamental physics.

Sitting in his office, Giorgio explained the focus of his research to me. I had questions. Given that time and space didn't exist before the "Big Bang"—a term coined by Catholic priest and astronomer Georges Lemaître to describe the universe originating from a single dense point[21]—I asked Giorgio if it made sense to discuss what came "before" the Big Bang. If "before" refers to events in time, how could there be a "before" if time itself didn't yet exist? Giorgio replied thoughtfully, "Yeah. It's quite difficult. We don't know. The answer is, we don't know. It's still a matter of investigation. What we know is that our universe comes from a hot phase, a very dense phenomenon we call the Big Bang. That's it. We don't know more than that." His admission highlighted how deep cognitive reflection on cosmic questions can lead to intellectual humility comparable to the humility triggered by direct experiences of space exploration.

Giorgio's understanding of the universe didn't come from direct observation in space but rather from decades of intensive study, mathematical analysis, and experiments. His humility emerged from recognizing the sheer complexity and mystery of the topics he studied. A similar dynamic became apparent when I interviewed an expert in space medicine I'll call "Quinn," whom I met at conference. Quinn had an impressive grasp of the biological challenges faced by humans in outer space. He envisioned a future in which humans might adapt to microgravity, creating societies fundamentally different from those on Earth. A devout Christian, Quinn viewed his medical and scientific work as distinct from his religious faith.

What particularly fascinated me about Quinn was his exploration of what could be learned not just from outer space but also from inner space—the subatomic realm. This shift in focus illuminated how Quinn navigated his scientific and religious worlds. When I asked Quinn if scientific insights ever influenced his religious beliefs, he became energized, enthusiastically discussing quantum theory. He suggested that the mysteries of quantum physics offered a unique perspective on what he described as "God's creation," precisely because at the subatomic level, "We ain't got no idea what's going on there!"

Quinn described how ordinary objects, such as the conference table at which we were sitting, appeared solid but were fundamentally made of empty space and probability functions. This scientific puzzle led him back to cosmological mysteries like dark matter and dark energy, highlighting how limited human understanding remains despite significant scientific advances.

Importantly, Quinn's viewpoint exemplifies how religious scientists might integrate their beliefs with their professional inquiries—not by claiming that their work objectively validates their religious perspectives but by finding meaningful resonance between their faith and their scientific curiosity. Quinn acknowledged the limits of human understanding, saying, "Can we really say that we fundamentally understand anything? What does it even mean? This calls for a great deal of humility, from the science side and from the religious side as well. . . . You follow any one of those paths too far, they seem to think that they have it figured out."

From an analytical standpoint, Quinn's perspective demonstrates how a religious scientist can use scientific uncertainty as a source of awe and wonder, rather than proof of religious truth. His criticism was directed at the overconfidence found in both scientific and religious communities, emphasizing that true scientific and religious inquiry must embrace uncertainty rather than deny it. Quinn's reflections thus provide a valuable insight into how individuals reconcile complex, sometimes competing frameworks for understanding the universe.

It seems that contemplating ideas or phenomena that are overwhelmingly vast, impressive, and difficult to fully understand often leads to an emotional blend of excitement, joy, dread, and fear—

experiences of awe, as discussed earlier in this book. Astronauts Tom and Cara, who encountered awe experientially by traveling to space, described directly feeling their own smallness compared to the universe's immensity. Scientists like Giorgio and Quinn, despite never leaving Earth, reached similar realizations cognitively, through sustained scientific inquiry into the universe. Both experiential and cognitive pathways thus lead to an awareness of human smallness and limited understanding, triggering intellectual humility.

Experiential awe, however, seems to serve as a shortcut to humility. You don't need an in-depth understanding of quantum mechanics or black holes to feel deeply humbled when directly confronting the immensity of space. Both the overview effect and the ultraview effect—experiencing awe by seeing Earth from outside it and contemplating the infinite cosmos beyond our planet—leave immediate and lasting impressions precisely because they abruptly reveal the inadequacy of familiar ways of explaining reality. When existing explanations and assumptions are suddenly insufficient, the gaps in our understanding become apparent. It's only through embracing this void—this sense of not knowing—that genuine curiosity and openness to new knowledge can flourish.

But humility isn't always welcomed or easy to embrace. Next we'll explore why some individuals resist humility, constructing emotional and cognitive barriers against awe to avoid confronting their own sense of smallness. Unfortunately, while such resistance might temporarily protect one's self-image, it also diminishes opportunities for wonder, limits personal growth, and negatively affects relationships and social interactions. Let's consider some approaches to accepting and using humility constructively and intentionally fostering awe in our lives.

5 *Navigating Humility*

Accepting Limits and Expanding Possibilities

Physicist Albert Einstein once observed, "As a human being, one has been endowed with just enough intelligence to be able to see clearly how utterly inadequate that intelligence is when confronted with what exists. If such humility could be conveyed to everybody, the world of human activities would be more appealing."[1] Einstein meant that our understanding of reality—ranging from the vast reaches of the universe to the microscopic world of subatomic particles—is severely limited. He worried that an inclination to pretend complete knowledge, refusing to admit how little humans truly understand, undermines our full potential. Admitting to limited understanding can initially feel uncomfortable, even unsettling. But the benefits far outweigh any temporary discomfort.

Although both awe and the humility it can inspire vary according to an individual's personal or cultural background, some intriguing universal patterns emerge. For instance, across cultures and historical periods, people have frequently struggled to accept the full extent of what they don't know, preferring instead the comfort of even illusory certainty. Genuine humility, however, often embraced through spiritual or religious frameworks, encourages individuals to accept and even cherish the unknown. Spiritual traditions repeatedly highlight the value of humility, especially when inspired by awe, as a deeply enriching and culturally embedded experience. Extraordinary encounters—such as the awe astronauts feel during the ultraview effect—can foster humility even among those who remain on Earth.

Humility versus Humiliation

To better understand the value of humility, it helps to clarify its meaning. Humility and humiliation are often confused, but they're distinct: humility grounds us, while humiliation shames us. Yet, as social psychologist Daryl Van Tongeren observes, these concepts sometimes blur. In Western culture, Van Tongeren notes, boasting or prideful behavior may provoke public humiliation, effectively serving as a social mechanism to enforce humility. This fear of humiliation, he argues, leads many people to downplay their accomplishments or deflect praise, producing a form of humility shaped more by the threat of shame than genuine modesty.[2]

While this humility-humiliation hybrid may act as a check on arrogance, it can also discourage individuals from pursuing opportunities or sharing successes that could benefit the broader community. An example of this is when talented or professionally accomplished women give up or downplay their own careers in order for their husband to be seen as the more successful spouse.[3] The fear that a woman's success could humiliate the family unit by "emasculating" her husband has caused some women to abandon work they enjoyed and that benefited society.

Van Tongeren identifies another connection linking humiliation and humility: Many people striving for success view humility as old-fashioned and restrictive, equating it with weakness and humiliation. He states, "Modern culture asserts that strength comes from power, aggression, and dominance. . . . So those who adhere to anachronistic notions of modesty . . . aren't going to get anywhere in life."[4] Individuals with this perspective believe career success demands projecting a facade of confidence, hiding vulnerabilities behind a carefully constructed image. This can be seen among politicians who pretend to have extensive knowledge about complex topics such as climate change, energy policy, military strategy, or epidemiology rather than admitting ignorance and deferring to genuine experts.

When we turn to individuals involved in space exploration, however, we encounter a somewhat different picture. While arrogance certainly exists in this field, the unique nature of working with space—the ever-present reality of confronting vast, unknown, and humbling cosmic scales—at least offers an environment conducive to fostering genuine humility. This form of humility expands possibilities by encouraging openness and acknowledging limits of knowledge. Those who embrace it can learn more and achieve greater success, precisely because they remain keenly aware of humanity's small place in the immense cosmos.

The humility gained by understanding humanity's true place in the universe, however, can be elusive. Unlike the humility shaped by parental discipline or societal bullying that Van Tongeren describes, it emerges from confronting the unknowable, from standing face-to-face with realities that defy human comprehension. Many prefer to avoid encounters with truths that disrupt our illusion of being at the universe's center. Yet it is exactly this kind of engagement that fosters the beneficial humility that is frequently evident among astronauts and others who explore outer space. "Carol," a member of the space community whose work focuses on robotic space exploration, exemplifies this form of humility.

Humility as a Virtue

Genuine humility isn't just about modesty—it's also a powerful virtue rooted in our interactions with unknowns. You probably won't be surprised to learn that I'm the kind of person who stays up late or gets up early to watch rockets take off or receive confirmation that a probe traveling for months in space reached its intended target safely. I love watching groups of engineers in mission control, whether at national space centers or private space companies, monitor the countdown; wait for signals; or verify that a probe, rover, or robot they sent to a faraway destination does what it's supposed to do. These teams are clearly nervous but also precise and careful, reminding me of a circus performer walking a tightrope. When it goes well, mission control becomes a different place. The once-silent team members rise to their feet, cheer, and hug. I'm always flooded with happy tears in those moments, joyful to see relief and pride wash over the exhausted experts.

I first saw Carol onscreen during one such triumphant event (I am unable to name the mission or where she works, as few women hold similar positions and I want to protect her identity). I was impressed by her tranquility in a room filled with anxious people. Even before the success of the mission was confirmed, she radiated calm. Some time later I saw her in the audience at a talk we both attended and introduced myself. She agreed to an interview, and soon I was experiencing her warm, unruffled demeanor firsthand.

After talking for a while, I began to think Carol might be a Quaker (an informal name for a member of the Christian denomination known as the Society of Friends) and she soon confirmed that fact. I wasn't sure what had led me to my guess—I had met other Quakers working in the American space program and that was probably part of it—but the calm, gentle manner with which she spoke suggested humility, a profound Quaker value.[5] George Fox, who founded the Society of Friends in the seventeenth century, saw humility as a great virtue, a quality that helped the Quakers withstand the early persecution that drove them from England to the American colonies.[6] Quaker humbleness influenced everything from their plain clothing to their use of the friendly and familiar "thou" with everyone regardless of their status, to their commitment to egalitarian worship.[7] When I told her she had confirmed my suspicions, she said she hoped that meant that her life was speaking those values. Carol's Quaker-flavored humility and collaborative spirit are particularly striking given the historically male-dominated context of the space community, where traditional gender roles have often marginalized or minimized women's contributions.

I suspect the sources of Carol's humility—her Quaker faith as well as her deep understanding of space—reinforced each other. After all, both religion and astronomy strive to comprehend mysteries beyond ordinary human experience. Someone like Carol, equally well-versed in spirituality and space exploration, would naturally view creation—and perhaps any Creator—as profoundly mysterious yet ultimately knowable through patient investigation.

Carol's childhood was marked by deep curiosity, especially about our planetary neighbors. She became increasingly inquiring as she realized the limits of adult knowledge, explaining, "You think that everybody knows everything and they're teaching you. But your exposure to the solar system, people *don't* know everything. There is so much we don't know. And I think that's what drew me. . . . That's the draw of space because you're *really* out there. You don't know anything." As Carol talked about her work, it was clear that although she and her team were expanding knowledge of outer space through robotic exploration, the more they learned, the more they understood how little they actually knew.

Carol's willingness to accept that she didn't know everything was part of her work in space but also part of her work with a team. She explained the diverse backgrounds and perspectives of her work "family" and her efforts to avoid assuming she already knew what they thought or were experiencing. She also stated that open communication about their differences allowed the team to work more smoothly. It was clear that for Carol, reminding herself about what she didn't actually know both prevented her from making mistakes based on faulty assumptions and also made her more receptive, another quality valued in her faith tradition.

There are no ministers in the Quaker faith and no hierarchy. Instead, every member of the congregation is equally able to lead when so moved by the Holy Spirit. A typical Quaker service involves congregants sitting together in silence until, as Carol explained, "somebody may stand up and speak. Sometimes the whole meeting there's no spoken word, but . . . when someone stands up and they speak the very thing that is in *your* mind, it's like this collective thought that somebody eventually speaks." Carol's description suggests a humility rooted in openness—an awareness, emphasized within Quaker tradition, that human understanding is incomplete and that insights might be guided by something beyond ourselves if we remain receptive.

This guidance, she thought, might extend to human work exploring space. Carol was careful in her phrasing: "I mean I don't believe we have some God up there that's forcing us to do something . . . but I do feel that these leaps in technology, these collective thoughts that are going on are influenced by something that is a higher entity, whatever that is." Even in her description of a higher power, Carol held to her conviction that the knowledge she had was limited, and she embraced that limitation. Both her career and her spirituality reinforced that conviction.

Living comfortably while maintaining awareness of one's own lack of understanding was something Carol could do, but it's not easy for everyone. Because of this challenge, many religions encourage the development of humility through specific practices such as the Christian traditions of foot-washing on Maundy Thursday or confession in Catholicism, *Tawadhu'* (humble conduct) in Islam, or even the custom among certain Hindu artists of not signing their artwork, all ways to acknowledge and embrace human limitations.

Humility within the space community, as exemplified by Carol, often draws on scientific understandings of the universe's vastness as well as deeply rooted cultural and religious ideas. Because astronauts, scientists, and engineers come from diverse backgrounds, including a variety of spiritual traditions, their perspectives on humility can reflect the religious and philosophical values they bring with them. Recognizing these connections can help us appreciate why humility remains an essential—and often profound—experience for those who explore space.

Religious Views on Humility

Religion has long emphasized humility as a key spiritual value across different faiths. The focus on humility in Carol's life arose not just from her scientific understanding of humanity's small place in the cosmos but also from the emphasis placed on humility as a spiritual virtue within her Christian denomination, the Society of Friends. In many Christian traditions, humility—the recognition and acceptance of one's limitations and an openness to the needs of others—is considered central to spiritual development. This value is illustrated vividly in biblical narratives such as Jesus washing the feet of his apostles (John 13:1-5), a profound symbol of humility and service, and echoed in teachings like Paul's warning in his letter to the Romans against thinking too highly of oneself (Rom. 12:3). Catholic tradition notably amplifies this emphasis, frequently quoting St. Augustine's assertion, "The way to Christ is first through humility, second through humility, third through humility."[8] Of course, humility is an ideal; religious adherents, like all humans, often struggle to balance spiritual humility with personal ambition and pride—a tension often visible in today's politically charged contexts.

Humility plays a significant role in religions beyond Christianity, including Judaism. The great prophet Moses is described in scripture as being notably "meek," an example that underscores the importance Judaism places on humility. Although interpretations of humility vary within Judaism's diverse traditions, it is broadly recognized as an essential religious and ethical value. Rabbi Sol Roth, an expert on Jewish ethics, emphasizes this point in his article "Towards a Definition of Humility," explaining, "Judaism requires humility on both religious and human grounds. It is a necessary ingredient in the religious perspective and is indispensable in social relations. The importance assigned to this character trait is considerable. In the Torah—the written and the oral—humility is associated prominently with Judaism's most outstanding representatives. No one was ever as humble as Moses."[9] Rabbi Roth's perspective highlights Judaism's distinctive emphasis on humility as spiritually meaningful and socially essential, promoting harmony rather than competition—a theme resonating with broader human reflections on our fragile place within the larger cosmos.

Islam, another Abrahamic religion, originated in Mecca (now in Saudi Arabia) at a time when both Judaism and Christianity were prominent in the Arabian Peninsula.[10] Christianity had arisen primarily from reinterpretations of Jewish scripture and prophecy. Islam, emerging later, developed in a context influenced by both Judaism and Christianity, with the Qur'an expanding upon and revising narratives from Jewish and Christian texts.[11] Given humility's prominence as a virtue in the other two Abrahamic religions, it is unsurprising that Islam also advises humility.

In Islam, as in Judaism, humility before God (*Allah* means "the Lord" or "the God" in Arabic) and in social settings are both important. The Qur'an states, for instance, "Successful indeed are the believers who humble themselves in their prayers" (23:1–2), but also advises, "And turn not your face away from men with pride, nor walk in insolence through the earth. Verily, God likes not each arrogant boaster" (31:18). The word *Islam* means submission (to God) in Arabic, and when submission is mentioned in daily Islamic prayers, the faithful press their foreheads to the ground, physically demonstrating humility before Allah. This symbolic act highlights the human response to encountering something vast and powerful, an experience present both within and beyond religious contexts.

Moving eastward to the two most prominent religions in India—Hinduism and Buddhism—humility is, again, a core spiritual practice, reinforcing our understanding of it as a human response to encountering the vast or overwhelming. Hinduism, a collective name for diverse religious practices sharing history and ideas, emphasizes concepts like *ahimsa* (nonviolence) and *dharma* (duty). Traditionally, duty has been associated with caste, linking profession, diet, and spiritual expectations to one's birth family and cultural group.[12] Both rights and responsibilities were thus linked to caste, creating a system in which arrogance might be seen as stepping out of bounds.

The story of Narada and Vishnu illustrates Hindu attitudes toward humility.[13] (Although some Hindus are monotheists, viewing all gods and goddesses as manifestations of a single divine entity, Brahma, Vishnu, and Shiva—the triumvirate—are frequently recognized as the religion's major gods.) In the Hindu scripture known as the *Shiva Purana*, a sage named Narada is intensely loyal to Vishnu. This loyalty deepens when, during intense meditation, Narada sees Vishnu. Vishnu praises Narada's devotion but declares Narada will not see him again until his death, when they will reunite.

Initially humbled by this recognition of his faithfulness, Narada soon encounters something that tests his humility. In Hinduism, devotion (*bhakti*) is a source of genuine power, so Narada's spiritual strength attracts the attention of Indra, king of the gods, who sees Narada as a threat. Indra attempts to divert him from meditation by sending the god of desire, Kamadeva, to distract him with temptations. Kamadeva fails to even find Narada due to an illusionary force (*maya*) placed by Shiva. Narada, unaware of this protective illusion, believes his own spiritual power stopped Kamadeva, fostering pride.

Seeing Narada's growing arrogance, Shiva, Vishnu, and Vishnu's consort Lakshmi extend the *maya*, tricking Narada into believing he has entered a beautiful city. There he meets a "princess" (actually Lakshmi) and, despite a vow of celibacy, desires her. Unaware it's an illusion, Narada asks Vishnu to enhance his attractiveness, only to have Vishnu give him a monkey's face. Initially cursing Vishnu in wounded pride, Narada eventually realizes his downfall stemmed from his lack of humility. The illusion dissolves, and Narada resumes devotion, humbled and wiser—a lesson on humility as the appropriate human response to realities beyond our understanding.

This Hindu story of Narada suggests that religious devotion requires humility, prefiguring insights from the Abrahamic religions and echoing a broader human truth: profound experiences often reveal human limitations and inspire humility. In Buddhism, humility takes a somewhat different form. When Siddhartha Gautama, a Hindu prince, underwent enlightenment and founded Buddhism in the fifth century BCE, he rejected many conventional Hindu beliefs, including the worship of gods. Leaving questions about the existence or importancc of deities unanswered, Siddhartha, known as the Buddha ("enlightened one"), argued instead that our perceived reality is an illusion, and the goal is to transcend this illusion, ceasing reincarnation and reaching Nirvana. Thus, the notion of gods became largely irrelevant, merely part of the illusion.[14]

The Buddha's enlightenment couldn't simply be transmitted to followers; individuals had to achieve it firsthand, guided by practices like meditation. After his death, Buddhism spread across Asia, developing various schools with the shared goal of enlightenment.[15] One approach, prominent in Japan's Zen Buddhist tradition, emphasizes a type of humility known as Shoshin, or "beginner's mind."

Beginner's Mind

Shunryu Suzuki's *Zen Mind, Beginner's Mind*, originally published in 1970, is one of the best-known texts on the subject. From a Zen perspective, enlightenment will come more easily if a person isn't burdened by conventional thinking. This is why Zen koans, unanswerable questions like "What is the sound of one hand clapping?," are used, and it's also why beginner's mind is so important. When confronted with a question that initially seems bizarre, a person cannot respond in an automatic way. Instead, the brain has to stop and process the question, starting from scratch. Rather than interacting with the world in an anesthetized or automatic manner, cultivating beginner's mind helps us genuinely notice and engage with the details around us.

At the beginning of his book, Suzuki writes, "The practice of Zen mind is beginner's mind. The innocence of the first inquiry—what am I?—is needed throughout Zen practice. The mind of the beginner is empty, free of the habits of the expert, ready to accept, to doubt, and open to all the possibilities." He paraphrases this same idea later, writing, "In the beginner's mind there are many possibilities, but in the expert's there are few."[16] Let's consider Suzuki's insight in light of the examples we've previously encountered, such as Zack gazing out at the Milky Way; Cody reflecting on the immensity of the universe; or Carol, whose humility is shaped both by her religious practice and her awareness of how much of outer space remains unknown. In all three cases we have individuals who know enough about the cosmos to be awed by it and whose awe reveals their "expertise" to be an illusion. Instead, they're brought back to Suzuki's question, "What am I?"

Psychologist Stephen Murphy-Shigematsu, who uses Zen approaches to teach premedical students, has argued that the essence of beginner's mind is vulnerability. He writes, "The experience of vulnerability helps to develop humility in a lifelong commitment to self-reflection rather than a detached mastery of a

FIGURE 7. A child experiencing awe during a meteor shower, illustrating how "beginner's mind" may be sparked by contemplating the cosmos. Illustration by Karl Tate.

finite body of knowledge. Vulnerability brings the lightness of beginner's mind rather than the heaviness of needing to be competent, appreciating mystery as much as mastery, listening more than speaking."[17] There are many paths to vulnerability, but when one contemplates the cosmos and feels awe, even that uncomfortable "shrinking" sensation Edmund Burke associated with exposure to the sublime, a sense of vulnerability occurs.

Perhaps it is the vulnerability required to be able to feel humility, however, that makes it so painful to some people. Psychologists have argued that awe can force something they call "accommodation," defined as "altering mental frames or schemas according to new incoming information. . . . Elements of novelty and surprise are also involved with this dimension."[18] In other words, when you feel awe, everything you thought you knew is suddenly called into question. Your confidence disappears, and your sense of reality is warped. Your security in your own power is diminished, and you feel humble. Evidence from psychology, anthropology, and other social sciences suggests these changes are positive, yet some people find the emotional exposure and vulnerability that often follow awe to be so uncomfortable that they'd rather avoid it altogether. Others respond to potentially awe-inspiring situations by rationalizing or explaining them away, perhaps without even realizing it, shielding themselves from awe and the disquieting sense of smallness it can inspire.

Overlooking Awe

It's common to shy away from or reject awe-inspiring experiences. Many people on Earth (and in space, for that matter) focus primarily on immediate priorities—families, jobs, chores, and hobbies.

While they develop expertise in their own areas, they rarely seek out experiences that overwhelm them with the realization of how little they actually know. Yet everyday encounters could easily inspire awe, from contemplating the bacterial content of raindrops to learning that colonies of living organisms survive harsh winters beneath our lawns, to reflecting on the chemical composition of the Sun. Even a quick conversation may remind us that the thoughts of the person next to us in the grocery line are just as complex and fascinating as our own. Acknowledging such truths requires confronting limited understanding and awareness, which may be uncomfortable.

At a young age, I was taught that thunder came from clouds crashing into one another. It seemed logical—I didn't know much about clouds and someone I trusted believed it, so I didn't question it. According to the National Weather Service, however, lightning rapidly heats the surrounding air to about 50,000 degrees Fahrenheit, a temperature roughly five times hotter than the surface of the Sun. The extreme heat causes air to expand explosively and then contract quickly as it cools, creating the sound wave we hear as thunder. Although accurate, this explanation is harder to intuitively grasp, as it involves understanding how gases behave under rapid temperature changes—something rarely encountered in daily life. Learning the truth forced me to acknowledge my ignorance and seek understanding, making thunderstorms far more fascinating to me.

Although it literally surrounds Earth, outer space remains easy to ignore. Many primarily regard the sky as the source of weather conditions. The Sun provides heat and light, appearing to move predictably, while clouds generate rain or snow, perhaps influencing moods and emotions. The Sun doesn't need to be thought of as a star; it can simply be seen as part of the sky, an extension of our world. The nighttime sky, however, can be harder to ignore. With little else visible, stars, planets, and occasionally dramatic events like comets, meteors, or supernovas may be revealed. Historically, careful observers noted these celestial phenomena, integrating them into myths, legends, and spiritual beliefs. Even today, a clear night sky can easily inspire awe.

Still, for many of us, it's entirely possible to move through life without experiencing that awe regularly. Practical reasons play a role—perhaps some rarely pause to look up, or urban environments limit our view—but there is also the tendency to underestimate or overlook the complexity and wonder around us. Whether it's assuming the simplicity of natural phenomena or undervaluing the rich inner lives of those around us, recognizing and challenging these assumptions might encourage a deeper appreciation of awe, making room for greater humility and openness to the unknown. There is, however, an additional factor worth considering: how our minds handle the tension between scientific knowledge and everyday experience.

When We Can't Believe Our Own Eyes

Our perceptions often contradict scientific realities, making certain truths difficult to fully internalize. Throughout this book, I've been describing our world as though it is a globe-shaped planet orbiting a medium-sized yellow dwarf star that itself is just a small part of the Milky Way galaxy. This observation, however, likely differs from your everyday experience. While the night sky can inspire wonder, the experiences of daily life give humans the impression that the Earth is fairly flat, with the sky above and the ground beneath our feet. It's easy to view the sky as a sort of overhead screen on which various objects—clouds, stars, planets, the Sun, and the Moon—move around and sometimes change shape. Thus, the true nature of our reality isn't obvious simply by looking around.

Physicists refer to the intuitive way humans understand reality before learning formal science as "naïve physics." Much of this is intuitive, rooted in observations and what feels like common sense. Research on children's naïve physics shows that they tend to assume if an object keeps moving, a force must be driving it. I'm not referring to the strike that sets a kickball in motion, for example, but rather what maintains the ball's movement afterward. It's easy to adopt the intuitive but false idea that the ball continues to move only as long as it retains some force from the kick. It's less obvious to grasp that an object in motion will remain in motion unless acted upon by an external force (Newton's First Law of Motion). Watching the kickball roll by, I'm probably not thinking about the friction and drag from the blades of grass beneath it or the air slowing it down as it passes. These external interferences are precisely what stop the ball. In contrast, if a ball were kicked in a friction-free environment like the vacuum of space, it would continue moving indefinitely or until another external force acted on it. This violates our firsthand experience on Earth and often comes as a surprise because it contradicts our expectations.[19]

Research into children's naïve physics suggests that how humans intuitively perceive the world often contrasts with how our brains logically process and understand it. Children initially grasp the world through physical interactions, forming a picture of reality based on sight, sound, and bodily sensations when kicking a ball, climbing a tree, or falling off a bicycle. Later, when introduced to formal physics through classes or textbooks, children must replace their intuitive (but incorrect) theories with new information grounded in logic, experimentation, and mathematical analysis.[20] The mechanisms behind these illusions are revealed, akin to a magician explaining how mirrors make a rabbit disappear.

Many children first encounter nonintuitive ideas at science fairs. Physics and chemistry exhibits attract crowds because it feels magical when what you predict differs dramatically from what actually occurs. Demonstrations of magnet-based levitation or remarkable chemical color changes often serve as young scientists' first experiences of awe, prompting accommodation and fundamentally altering their understanding of reality. Some children, however, bypass the physics and chemistry tables to examine the turtles at the zoology exhibit instead. They might never receive the message that their naïve understanding of the world is flawed. I'm proof that a student can navigate public education, college, and even graduate school without learning fundamental physics concepts—though fortunately, friends and mentors later helped me fill in those gaps in knowledge.

Reality through a Cultural Lens

Both intuition and cultural frameworks influence how reality is interpreted. The intuitive understanding children seem to have of the world is one reason why the facts about the universe can seem difficult to absorb, but culture also plays a role. Culture—the understandings, activities, and ways of interpreting reality that are shared within a group and passed down to children—is our main source of information about the world. It could be argued that culture influences our worldview even more intensely than our physical experiences or the scientific data. The explanations of reality I receive from my family, group, tribe, or country don't just give me an understanding of the world, they also provide me with an identity. My sense of self might not be put at risk by learning the real cause of thunder, but it might feel threatening to discover that the universe is much different than what I've been taught by people I trust.

An example from a particular society can help illustrate how this works. The Yolngu (or Yolŋu) people of northeast Arnhem Land, in northern Australia, understand the Milky Way as a river of stars inhabited by their ancestors' spirits. This river forms the pathway connecting the realm of ancestors with our living space on Earth, allowing the powerful beings who exist in the sky to connect with and influence our lives. This cosmological view is central to Yolngu identity, informing the explanations of why people exist, their purpose on Earth, and what will happen in the future. This worldview represents a meaningful way of understanding the cosmos, independent of Western scientific models. It does not, however, prevent Yolngu community members from engaging with science. In fact, Charles Darwin University in Darwin, Australia, has established the Yolngu Aboriginal Consultants Initiative, involving Yolngu people in research and valuing their insights by connecting these perspectives with Western science.

The Yolngu understanding of the sky also interests anthropologists who study cultures worldwide to document shared interpretations of the cosmos. Researchers called "cultural astronomers" collect folk beliefs and stories to reconstruct how ancient people experienced the heavens. Yolngu narratives about constellations such as Orion and the Pleiades aren't just fanciful tales—they reflect real astronomical changes over thousands of years. For instance, the Pleiades, also called the Seven Sisters, currently only has six easily visible stars. Ancient stories, including those from

Indigenous Australian communities, frequently explain why the seventh sister-star is no longer visible. These stories provide evidence that the constellation mattered to humans when all seven stars could be seen. Such narratives have been passed through generations, often shared among various groups. Similar stories among Australian populations like the Yolngu and in ancient Greek texts help us better understand human migration patterns and cultural interactions over time.[21]

Culture can also combine with naïve physics, along with the human need for community, to make the Earth's "flatness" seem intuitive, giving rise to the "flat Earth" theory. Contemporary flat Earthers, despite abundant contradictory evidence, argue that the Earth is shaped more like a disk than an orb. They claim that the well-supported view of Earth as globe-shaped is a lie imposed on a gullible public.[22] Interestingly, while some ancient populations initially believed in a flat Earth and later recognized its roundness as their knowledge and technology advanced, today's flat Earth movement isn't a continuation of those ancient beliefs. Instead, it's a deliberate rejection of scientific concepts, usually by individuals raised within societies that have evidence-based understandings of Earth, the solar system, and the universe.

Researchers suggest that the recent popularity of flat Earth belief is tied more to social identity than evidence. It gives adherents a sense of community united against perceived powerful forces insisting reality differs from immediate perception. For flat Earthers to maintain their perspective and group affiliation, they often employ "motivated ignorance," deliberately avoiding information that could threaten their social identity.[23] This requires viewing governments, scientists, astronauts, physicists, and astronomers as conspirators seeking to hide Earth's true shape from the public. Today's flat Earth movement traces directly to the nineteenth-century English writer Samuel Rowbotham, who published a tract titled "Zetetic Astronomy: Earth Not a Globe" in 1849.[24] His work was revived by Samuel Shenton in 1956, leading to the establishment of the International Flat Earth Research Society (the Flat Earth Society) based in Dover, England. The movement remained small until it gained momentum with the rise of the internet.

Anthropologists define "culture" as beliefs and behaviors shared by groups and passed down to new generations or members. This clearly applies to the flat Earth movement, which involves heavy group interaction. With the widespread availability of the internet and social media, people can now connect more easily than ever, allowing groups like the Flat Earth Society to grow unconstrained by geography or social class. If you have an idea, you can likely find someone who shares that idea within minutes online. Flat Earth meetings and online discussions are enjoyable social events, offering participants a sense of superiority gained from believing they possess knowledge unavailable to the general public. Some members may be involved primarily as an act of rebellion against a society they distrust, even if they're not fully convinced of the Earth's flatness. However, research conducted at a flat Earth conference revealed a strong communal atmosphere, with speakers effectively reinforcing group beliefs.[25]

Believing in a flat Earth isn't logical or rational, but it doesn't need to be in order to fulfill a social function. It gives flat Earthers a sense of belonging to an exclusive community with privileged knowledge about reality. Anthropologists note similar patterns among religions with nonmainstream beliefs: The more ridicule or external persuasion adherents face, the stronger their attachment becomes. This pattern also characterizes other identity groups skeptical of science-based knowledge, such as communities rejecting vaccines or disputing the scientific consensus on climate change. To these groups, the outside world offers little; they're operating according to different rules.

Yet humility requires one to stay open to listening, even to ideas that are unusual or unpopular. It's beneficial to understand why someone believes the Earth is flat or why they distrust scientific consensus. At the same time, humility also means acknowledging the limits of our personal expertise and scrutinizing claims with available evidence. Ideally, this involves consulting experts who have spent their careers critically examining these subjects. For nonspecialists, distinguishing legitimate claims from misinformation is challenging. Part of staying open-minded involves recognizing this difficulty, being cautious about our assumptions, and thoughtfully evaluating the credibility of our sources. In other words, genuine humility involves both a willingness to listen and a commitment to careful, informed evaluation.

Staying Open

Ultimately, humility opens us to exploration, inspiring curiosity and driving us toward greater understanding. Apollo 14 Lunar Module Pilot Edgar Mitchell was on his way back from the Moon. Mitchell, who was an open-minded person and had even done surreptitious experiments into extrasensory perception during his lunar mission, suddenly had an experience he did not expect. He felt what he described as "an extraordinary, personal connectedness" with the heavens around him, explaining it as "an ecstasy of unity," overwhelmed by the sensation of physically and mentally extending out into the cosmos. Mitchell, who held a doctorate in science from the Massachusetts Institute of Technology in aerospace and astronautics, recognized that his strange feelings could have been the result of the space journey or the rotation of his spacecraft. He believed very strongly, however, that he had been changed by the experience. He described it as "an upwelling of fresh insight, coupled with a feeling of ubiquitous harmony—a sense of interconnectedness with the celestial bodies surrounding our spacecraft."[26]

This "ecstasy" wasn't Mitchell's only overwhelming experience in space. Like astronauts before and after him, he was profoundly affected by the vision of our planet: a beautiful, glowing sphere spinning in nothing. Like all of the astronauts I've spoken to, for him, seeing the Earth from orbit was more than sufficient proof that it was round, not flat. But the meaning of it, the deeper understanding, was sudden, overwhelming, and undeniable. In a 1974 interview with People magazine, Mitchell described seeing Earth from space this way: "You develop an instant global consciousness, a people orientation, an intense dissatisfaction with the state of the world, and a compulsion to do something about it. From out there on the moon, international politics looks so petty. You want to grab a politician by the scruff of the neck and drag him a quarter of a million miles out and say, 'Look at that, you son of a bitch.'"[27]

Mitchell's description captures something profound about experiences like the overview effect—and, by extension, the ultraview effect. Seeing Earth as a unified globe against the vastness of space often triggers a powerful shift in perspective, creating what Mitchell termed an "instant global consciousness" and a "people orientation." The sudden realization of how many human beings share our fragile planet fundamentally alters one's perspective.

Mitchell's frustration with political leaders who fail to grasp the bigger picture illustrates how strongly the overview effect can challenge self-centered views. His "compulsion to do something" also highlights how extraordinary experiences don't merely humble individuals, they also motivate them, inspiring a deeper commitment to collective action and responsibility.

Many astronauts who journey into space, whether undergoing the overview effect described eloquently by Mitchell or experiencing the ultraview effect described by some of my interlocutors, encounter feelings of awe at some point during their mission. Awe typically leads to humility, but humility isn't the end of the process. Recall a time when you learned something that shattered what you thought you knew. That realization might have made you feel small, but it probably also made you intensely curious. Edgar Mitchell, transformed by the awe evoked by traveling through space and viewing Earth from afar, found himself motivated by everything he recognized he didn't understand. He wanted both to improve the world and to understand precisely what that cosmic experience had meant.

Mitchell began exploring mysticism and "noetic" knowledge—information that enters the mind directly and instantly, rather than through normal processes of learning. He eventually founded the Institute of Noetic Sciences in Novato, California. For Mitchell and many others, experiences of awe followed by humility naturally lead to openness, a willingness and desire to actively explore new ideas and realms of understanding. This progression reflects insights from religious traditions discussed earlier, from the humility emphasized in Abrahamic faiths to the beginner's mind in Buddhism, in which humility opens the door to deeper understanding.

Edgar Mitchell embodied both scientific rigor and spiritual curiosity, exemplifying how qualities such as openness and curiosity—qualities deeply rooted in humility—are essential to meaningful scientific inquiry. Scientists form hypotheses, test their assumptions, and remain willing to abandon even cherished theories when evidence contradicts them, precisely because they recognize that humans do not fully understand the world. Scientific breakthroughs depend on researchers' willingness to experience the discomfort of uncertainty and admit the limits of existing knowledge. Thus, humility isn't merely a desirable personal virtue; it's a fundamental principle driving scientific exploration and advancement. As we move into the next chapter, we'll explore in greater depth how humility—grounded in awe and openness—not only inspires scientists like Mitchell but also provides the essential foundation for all kinds of exploration, from Earth's oceans and mountain peaks to the deepest reaches of space.

III *Exploring the Unknown*

6 *From Awe to Action*

Understanding Exploration

What does it mean to be "open"? Rosalia de Castro, a poet and novelist born in the pilgrimage center of Santiago de Compostela, Spain, once said, "I see my path, but I don't know where it leads. Not knowing where I'm going is what inspires me to travel it."[1] De Castro's words exemplify a particular reaction to embracing one's sense of "not knowing." That openness, that "beginner's mind," can create a real sense of adventure, grounded in a willingness to change one's mind and circumstances. In the previous chapter I examined how humility can come from experiencing something awe-inspiring: a huge building; a vast landscape; or the Milky Way galaxy, unimaginably dense with stars. True humility may inspire you to reshape yourself, sparking an intense desire to acquire knowledge or skills you realize you lack. When people think they know all there is to know, conversely, they often close themselves off from new information. When they recognize that there is still a lot to learn, however, they are encouraged to maintain an open frame of mind and be willing to expose themselves to new ideas, thoughts, or capacities.

In a very real way, then, an experience of awe—such as witnessing the enormity of the cosmos or the magnificence of nature—can start a chain reaction: awe leads to humbleness, and humility, if embraced, can bring about curiosity and an eagerness to explore. Embracing humility is like creating a vacuum, a truly empty space. They say "nature abhors a vacuum," meaning that it's not natural for a space, at least on planet Earth, to be unfilled by something, though there are exceptions.[2] A humble mind is not a vacuum, but a recognition of how little we know seems to make room for more information, new ideas, or experimentation—this is the trait known as "openness."

Research psychologist Robert R. McCrae, who developed the influential Five-Factor Model of personality, has explored openness (or "openness to experience") throughout his career, defining it as something "seen in the breadth, depth, and permeability of consciousness, and in the recurrent need to enlarge and examine experience."[3] Breaking this down, McCrae ties openness to human thought processes: someone who is broad-minded, reflective, and easily integrates new ideas—what he calls "permeability of consciousness"—can be described as open. He further argues that open people actively seek novelty and try to understand new experiences, demonstrating a strong urge to explore.

Some researchers, however, have questioned McCrae's Five-Factor Model, particularly its biological assumptions and the idea that personality traits remain constant.[4] Others suggest that this model, developed largely in Western contexts, might overlook cultural variations in personality.[5] Despite these criticisms, McCrae's description of openness provides a useful starting point for investigating the cycle of awe, humility, and exploration described by many astronauts. Awe arises from encountering something vast or profound, humility involves recognizing one's smallness within the grand scheme, and exploration is ultimately about growth—

encountering immensity inspires us to expand our own understanding. Later in this chapter, we'll examine ethical complexities associated with exploration, highlighting why the best examples are guided by responsibility and critical reflection. But first, let's see how awe-inspired openness shapes the life of "Sabine," a European space scientist and one of her nation's first female astronaut candidates. Sabine's story illustrates how an experience of awe—on Earth or in space—can inspire extraordinary humility, curiosity, and a powerful desire to explore.

Sabine

The idea that outer space can awe us, make us realize how small Earth is in comparison to the cosmos, and then inspire curiosity and a desire to grow our knowledge can happen in space, but similar things can happen on Earth. You may recall in Chapter One that on a stargazing expedition, my young son was suddenly full of questions, realizing he couldn't explain what he was seeing in the sky. A similar thing happened to Sabine.

Unlike some space community members I've interviewed, Sabine was not particularly religious. As a child she had shown some interest in her grandparents' religious beliefs but had eventually decided religion wasn't for her. That said, her entire career was inspired by the awe she had experienced looking into space. In her story, we can see how that initial feeling of wonder transformed into strong curiosity. During our conversation, Sabine relayed a childhood memory:

> One of the earliest moments where I realized I wanted to be a physicist, was lying down on the grass in the garden in the summertime.

> We wanted to watch the falling stars and what I realized looking out there, staring at the stars—there's no *ceiling* and what I'm looking at is *right into space*. And I'm looking right at these stars that are so *far away,* and it seemed like suddenly I could, you know . . .

Eyes wide, with a joyful grin, Sabine lifted her hands, palm side down, into the air, as if the little girl she had been began to levitate toward the sky. She continued:

> In our everyday world we have basically 2-D vision because we live on a flat surface. And what we see is maybe [just] a *little* bit up, you know? But not too much. And that sort of added the third dimension for me. I could *feel* that the Earth is a sphere and it's not actually flat. And I could see that there's no *ceiling* and the longer I looked, [the more] I looked deeper, deeper into space.

Sabine's awe of the stars and her realization that the sky had no "ceiling" transformed her, at age twelve, as much as any first look at our planet from space did for an astronaut. Sabine felt like she could fly off the planet. She understood the reality of outer space; she perceived the Earth's spherical shape. Unlike so many earthlings who never consider the sky at all or see the Moon and stars as little more than romantic decorations, Sabine *got it.* Her mind opened and the universe flooded in. The immensity of outer space overwhelmed and delighted her, making her realize just how little she understood and how much more there was to learn. The resulting curiosity fashioned young Sabine into an explorer and set her on a path to study space and perhaps go there herself. Her awe-fueled curiosity, then, exemplifies how humility inspired by cosmic wonder can underpin ethical and responsible exploration.

Sabine's vision of humanity's future in space demonstrates her enthusiasm and dedication to finding new knowledge:

> So what I would want is that we make some great technological advances and explore where we are in our galaxy . . . first, where Earth is in our solar system. Then we could travel to the moons and see if there is also life . . . and explore our realm. . . . But beyond that, space is so empty that I think it would take much more understanding. . . . I would hope for *that* in the next decades. . . . The question is what will happen after that? I mean, I would really hope that we can explore at least our own solar system . . . for example, travel to Saturn or Jupiter, as that will take more than a human lifetime to do.

Sabine is passionate about exploration. For Sabine and many others in the space community, exploration is the natural response to a powerful curiosity about the unknown, sparked by experiences of awe that vividly reveal just how much remains to be learned.

Yet we don't have to look exclusively to the stars to find examples of places and phenomena that humans don't fully comprehend. Indeed, exploration takes many forms. It happens every day—in laboratories, on mountains, in caves, underwater, deep underground, and even within the living bodies of animals, plants, and fungi. And I'm only scratching the surface. Such a wide range of phenomena shows how broadly and diversely humans feel compelled to investigate unfamiliar things. But what exactly is exploration? What makes us want to explore in the first place?

Exploring Exploration

While awe and humility have specific definitions, exploration is harder to pin down. We can explore by setting sail on an unknown sea, traveling to a new environment, trying a new cuisine, researching an unfamiliar topic, and more. Definitions include something to "investigate, study, or analyze," "become familiar with by testing or experimenting," "travel over (new territory) for adventure or discovery," and "examine especially for diagnostic purposes."[6] Yet none fully captures the "spirit of exploration."

How might an organization devoted to exploration define it? The Explorers Club—of which I'm a Fellow—is an international association that has been dedicated to exploration for more than a century. Originally named the Arctic Club, it was founded in 1904 during the fiercely competitive race to the North and South Poles, a period deeply intertwined with colonial ambitions and problematic ideologies of conquest. Recognizing this history is essential to appreciating exploration's ethical complexities.

The club's website invites new members into a community that includes "leaders in polar exploration, diving, aerospace exploration, archaeology, zoology, physics, oceanography, astronomy, ecology, geology, paleontology, conservation, mountaineering, and speleology." While comprehensive, the list isn't definitive. Membership criteria state that a "Member" must have "contributed broadly to the cause of exploration," while a "Fellow" must have made "actual contributions to scientific knowledge in geographical exploration or allied sciences." Clearly, however, the urge to explore isn't limited to formal criteria. Many people feel a compelling urge to discover and understand the unknown without formal recognition.

While the Explorers Club primarily recognizes professionals, exploration isn't limited to experts or formal expeditions. For our purposes, let's define exploration as actively seeking and engaging with the unknown, driven by curiosity and the desire for greater understanding, regardless of professional status. Astronauts orbiting the Moon and engineers working with robotic rovers on Mars are explorers; so too is someone learning a new language, experimenting with an unfamiliar art form, or planting an unusual crop in their garden. Each instance requires openness to uncertainty and curiosity—our next topic.

Curiosity

It is impossible to discuss exploration without talking about curiosity—the drive, found in nearly all humans, to learn more about something unfamiliar or not understood. It is the reason children ask so many questions, like when my son Luke, around two years old, pointed to a birthmark on my eyelid, asked me what it was, and after hearing my explanation confidently announced, "People like eye birthmarks!" He was clearly absorbing new information and integrating it into his growing understanding of the world.

Humans, of course, aren't the only beings to experience curiosity. Scientists studying animal curiosity refer to this trait as neophilia, an interest in or love of new things. Many readers may be familiar with the expression "curiosity killed the cat"—often with the addendum, "satisfaction brought him back." Domesticated pets are often incredibly curious, looking out windows, sniffing unfamiliar objects, or tilting their heads and maintaining eye contact when something inexplicable happens. In animals, as in humans, curiosity underpins the urge to explore.

Historically, curiosity—implicit in the artistic experimentation and innovative tool-making of early humans and related species—was likely one motivation behind the great migrations out of Africa by certain *Homo erectus* groups.[7] Scholars have suggested multiple explanations for these movements, including food availability, climate change, and following animal herds. Curiosity surely played a role too, at least indirectly, as exploring new territories inherently involves engaging with unfamiliar environments.

We cannot know precisely how early humans experienced curiosity emotionally or cognitively, but it's reasonable to infer that a species creating art, modifying tools, adapting to fluctuating environments, and devising music possessed some level of curiosity. Curiosity, as understood today, carries an optimistic perspective, involving the expectation that exploration will yield something new or valuable. Such openness, like tentatively tasting unfamiliar food, allows experiences to unfold rather than remain unexplored. While we can't confirm this perspective in our ancestors, curiosity's evident presence throughout human behavior and culture strongly suggests it is part of our shared heritage.

Not all living things or even all people are curious, however. The opposite of neophilia is neophobia, the tendency to avoid new things and stick to familiar patterns. Neophobia can be a lifesaver. Even though young children are usually very curious, they often become picky eaters between the ages of two and six. Scientists hypothesize that this pickiness results from children reaching an age when they can wander far from supervision, and curiosity about food could lead them to ingest poisonous berries or deadly mushrooms.[8] Neophobia, then, might be an adaptation, making children wary of unfamiliar foods. Keep this in mind the next time a child you know won't eat anything but grilled cheese sandwiches.

While neophobia clearly offers the benefit of safety, excessive caution can prevent us from considering new information that might solve problems or improve our understanding of the world. Cultivating a balanced degree of neophilia—a cautious openness to novelty—is likely essential for scientific or exploratory approaches. Curiosity varies significantly across individuals, cultures, and historical contexts, shaped by a complex interplay of biology, culture, and environment. Although not everyone experiences or expresses curiosity in the same way—or at all—fostering curiosity is often beneficial. An open-minded willingness to consider new ideas and information, when encouraged, supports our capacity to evaluate evidence fairly and objectively, as we will explore next.

Gathering Data with an Open Mind

How difficult is it to be objective and open-minded when your beliefs are challenged? Imagine your best friend is on trial for auto theft. You know this person well and you don't believe that they would ever commit such a crime. As you sit in the courtroom and the evidence mounts, would you remain loyal and steadfast, or would you allow yourself to consider the exhibits being presented by the prosecution? The situation demands that you choose between two traits that are generally positive: faithfulness, which compels you to stick by your friend, or objectivity, which requires you to be open to persuasion as long as the evidence is sound. This conflict between defining truth based on community membership, tradition, and fidelity or, instead, things that can be proved, without taking emotions or relationships into account, is at the heart of the dilemma that led to the scientific method.

The scientific method traces its origins to Socrates, specifically a characteristic that his student Plato described as "Socratic ignorance." In the *Dialogues of Plato*, written in the fourth century BCE, the philosopher describes situations in which Socrates was open about his own ignorance. Socrates, usually, believed that being humble about one's knowledge, particularly striving to be aware of both the strengths and limitations of one's understanding, was the wisest way to go through life.[9] This perspective also fosters critical thinking by allowing new ideas to be evaluated and debated, and encourages lifelong learning because of the principle that limited knowledge can always be improved.[10] In the trial situation described above, holding the perspective of Socratic ignorance would make you more likely to listen to evidence with an open mind, even if it pointed to your friend's guilt.

Socratic ignorance had a large influence on Western culture, especially during the Renaissance (1450–1600), when awareness of ignorance, self-knowledge, and intellectual humility were all cultivated in the pursuit of knowledge. During the middle of the fifteenth century, Catholic priest Marsilio Ficino translated Plato's complete works into Latin after being commissioned for the task by Cosimo de' Medici. Plato's philosophical ideas, as well as those of his mentor, Socrates, became much more widely known. Concepts related to Socratic ignorance, like "refutation," which is aimed at disproving false beliefs, and the "dialectical method," which uses careful analysis to determine the hidden qualities of things, were newly discovered tools in the pursuit of true knowledge.[11] These ideas found in Plato influenced such astronomers as Galileo Galilei and Nicolaus Copernicus, encouraging them, perhaps, to question and argue against the belief, largely accepted at the time, that the Sun orbited the Earth.[12] Loyalty to religious and

TABLE 1. Applying the scientific method versus reasoning based on belief or loyalty

	SCIENTIFIC	NOT SCIENTIFIC
Observation:	A car was reported stolen	"My friend would never steal"
Information gathering:	Examine fingerprints, alibi, video	Ask mutual friends if they trust suspect
Hypothesis:	Suspect could be guilty or innocent	"He's innocent because he's my friend"
Testing and prediction:	Objectively assess each piece of evidence	Remain convinced of innocence
Analysis and conclusion:	Accept evidence-based conclusion	Ignore conflicting evidence
Share and document:	Share transparent results with others	"Defend my friend no matter what"

Note: This table illustrates the differences between applying the scientific method versus relying on belief or loyalty. Using the scenario of evaluating whether a friend stole a car, it contrasts objective information gathering, hypothesis formation, testing, analysis, and transparent reporting (scientific) with subjective reasoning rooted in personal beliefs and loyalty (not scientific). Table by Karl Tate.

cultural ideas was being tested, and often rejected, as the Scientific Revolution, which took place between the mid-sixteenth and the late eighteenth centuries, began.

The scientific method is a systematic approach to gaining knowledge. At its core, it involves clearly defining the question to be answered, gathering measurable evidence through careful observation and experimentation, and rigorously testing the initial assumptions or hypotheses against that evidence. If this method were applied to your friend's trial, for example, it wouldn't be enough for the jury to hear testimony about his good character from family and friends. The prosecution would try to prove guilt based on data like his fingerprints on the steering wheel of the vehicle; his lack of an alibi for the evening the car was stolen; and his appearance on security cameras, showing him climbing into the car and driving it away. The scientific method says that if your heart tells you one thing and evidence something else, you have to believe the evidence, no matter how difficult. Interpreting data with an open mind means being humble enough to put aside what you *think* you know to consider and be swayed by good evidence.

The Historical Development of the Scientific Method

The scientific method wasn't invented by a single individual at one specific time or place. Instead, it emerged gradually over centuries, drawn from diverse forms of knowledge production around the world. In what is now Iraq, the tenth- and eleventh-century mathematician and scientist Ibn al-Haytham developed experimental techniques that clearly influenced European scholars of his era and later periods. The methods of al-Haytham and other scholars of the Islamic Golden Age significantly shaped the development of the scientific method.[13] During the Renaissance, scholars looked back to classical Greek thinkers such as Socrates for foundational ideas and methods. Later, in seventeenth-century Europe, new influences appeared—like William Gilbert, an English physician and philosopher whose *De Magnete* (*On the Magnet*), published in 1600, is regarded as one of the earliest scientific works grounded almost entirely in careful experimentation and observation.[14]

Gilbert sought to explain magnets and magnetism. Rather than choosing to simply accept his era's ideas about magnets (these explanations tended to be pretty mystical, tied to the movements of the stars or the "living soul" of iron), he used various instruments and experimental devices to gain data that he could analyze objectively. It wasn't enough for Gilbert to read conventional explanations of magnetism—he needed to see what was going on himself by performing carefully organized and recorded tests that could be replicated by others. Gilbert's work was published during the earlier period of the Scientific Revolution.

The Scientific Revolution is usually dated to 1543, with the publication of Nicolaus Copernicus's *De revolutionibus orbium coelestium (On the Revolutions of the Heavenly Spheres)*. As this new approach to knowledge took hold, the scientific method became the main way educated Europeans understood the world, making explanations found in ancient texts and religious scriptures seem less convincing. Scientists began to keep track of their observations and uncover patterns, leading to principles like Isaac Newton's laws of motion and Johannes Kepler's laws of planetary motion.[15] The world became less mysterious and better understood, even easier to control, eventually enabling large-scale transformations, like those of the Industrial Revolution in the eighteenth and nineteenth centuries. However, increased understanding and control were not always ethically neutral or universally beneficial. The Scientific Revolution coincided with and influenced another era in human history: the Age of Exploration. Despite significant discoveries, this era also brought colonization, exploitation, and lasting injustices—an ethical legacy that can't be ignored when it is examined today.

A Darker Side to Curiosity

The Age of Exploration, or Age of Discovery, spanned the 1400s through the 1700s, when European countries including England, Spain, France, Portugal, and the Netherlands began visiting and colonizing previously little-known parts of the world. These terms, however, romanticize a period marked by tragic loss and cultural upheaval for populations in the so-called newly discovered regions. Because these colonial powers had Atlantic access and the technology to build large sailing ships, it was relatively straightforward for them to reach distant lands. Although I've highlighted curiosity as a motivating factor, historically exploration also had a darker side. Beyond genuine interest in the unknown, explorers were strongly driven by desires for gold and valuable goods (like spices), territorial conquest, and religious conversion.

Moreover, European curiosity about distant places was deliberately amplified to secure political and financial backing for expeditions. Diganta Bhattacharya notes that expedition journals from this period intentionally intrigued readers, vividly portraying adventures and discoveries abroad. Genuine humility, however, was largely absent. Sometimes called the Age of European Expansion, this era followed the Enlightenment, when science flourished and systematic data collection became widespread. Rapidly increasing knowledge and global influence led many Europeans to feel entitled to claim whatever they desired. Populations in Europe often displayed excitement and a sense that the world was fully knowable, believing unprecedented understanding was within reach. Even though Europeans acknowledged the need for continued study, many maintained a persistent certainty about their superiority over encountered populations, believing they had a right to control these territories. Narrators of voyage accounts strategically emphasized dangers faced by explorers, immense oceanic challenges, and mysterious peoples, fostering addictive excitement among European readers.[16]

Today, people accustomed to a thoroughly mapped world—with geographic features precisely recorded by satellites and high-tech systems—may find it difficult to imagine the uncertainties early European explorers faced. Historical narratives often described encounters with unfamiliar societies' foods, clothing, and cultures in ways intended to fascinate readers. These portrayals cultivated a craving for exotic goods such as tomatoes, chocolate, potatoes, and tobacco—items that quickly became essential within European societies. While these exploration stories emphasized adventure and discovery, they frequently obscured the significant human costs involved. The excitement and curiosity promoted by these accounts reinforced colonial attitudes of superiority and entitlement, bringing destruction and disruption to Indigenous communities—an ethical reality that must be confronted. Understanding exploration's full consequences requires examining its ethical implications, which we'll consider next.

Two Models of Pacific Exploration

Global exploration was not, of course, limited to Europeans during the Age of Exploration. The Polynesian peoples, for instance, undertook extensive exploration and settlement of the Pacific Ocean using a practice known today as wayfinding. Some scientists suggest that these Austronesian-speaking peoples originally came from the region now known as Taiwan.[17] Over centuries, Polynesian navigators developed careful and complex methods for traversing the ocean, skillfully settling islands throughout the Pacific, including Samoa, Tonga, Fiji, and eventually the more distant Marquesas Islands, Rapa Nui (Easter Island), and Hawaiʻi. Remarkably, most of these islands were uninhabited, so Polynesians were generally able to expand without disrupting existing communities. Their method of wayfinding relied on sophisticated, careful observation of the stars, currents, winds, marine life, and subtle environmental cues, representing exploration driven by humility, openness to new information, and curiosity about their surroundings.[18]

Patrick V. Kirch, an anthropologist and expert on Polynesian wayfinding techniques, explains that Polynesian navigators relied on observations of the Sun, Moon, and stars, but also carefully attended to natural phenomena such as the flight patterns of birds, ocean swells, cloud shapes, and vegetation drifting in the water to locate islands—often across thousands of kilometers of open ocean. Kirch emphasizes that successful wayfinding involved techniques perfected and refined over generations, transmitted through oral traditions and hands-on training in an apprenticeship-style system. Like European navigators of later centuries, Polynesians were meticulous observers of the environment, accumulating detailed, practical knowledge about the natural world. What sets Polynesian wayfinding apart, however, was a culturally embedded attentiveness to uncertainty and environmental constraints. Polynesian navigators conducted shorter trial voyages to test routes and refine methods before undertaking ambitious journeys, a practice reflected archaeologically in settlement histories characterized by long pauses and periods of careful adaptation. While not indicative of inherent moral superiority, these deliberate pauses and careful preparations underscore a cultural awareness of environmental constraints and the traits of humility and openness toward the unknown.[19]

While European explorers during this period similarly depended on meticulous environmental observations and detailed navigation skills, their explorations unfolded within vastly different social, political, and cultural contexts. British naturalist Joseph Banks, who traveled with Captain James Cook aboard the HMS

Endeavour, provides an illustrative example. Banks systematically documented unknown plants, animals, and societies, laying groundwork for future scientific studies.[20] Yet his behavior also reflected problematic colonial attitudes common in that era; historians critically note that Banks "treated Tahiti as a free love zone," potentially causing disease and harm to Indigenous peoples.[21] Banks's complicated legacy highlights a central ethical tension in exploration: curiosity and careful observation, when combined with entitlement and the assumption of cultural superiority, can lead to ethically compromised and exploitative practices.[22] Rather than humility, a sense of certainty in cultural superiority shaped many European interactions with Indigenous populations, often leading to disruption and causing lasting damage.

Captain Cook's voyages similarly demonstrate how a lack of cultural humility and openness can lead to unintended and tragic consequences. During his third expedition, Cook's initial interactions with Hawaiians, while marked by a dramatic power differential, were relatively respectful and characterized by mutual curiosity. Yet when tensions arose—largely due to the crew's attempts to assert control based on their own cultural expectations—Cook quickly resorted to force. In an effort to recover property he believed had been stolen, Cook attempted to take a Hawaiian chief hostage, escalating tensions. This attempt to enforce European norms in an unfamiliar society illustrates how a lack of humility and openness in exploration can have bad outcomes. The resulting conflict led directly to Cook's death, underscoring how ethical complexities in exploration are deeply tied to explorers' attitudes, assumptions, and willingness (or unwillingness) to respect local ways of life.[23]

Polynesian wayfinding and European colonial voyages offer contrasting yet nuanced lessons about ethics in exploration.

Genuine curiosity paired with humility and careful observation, as seen in Polynesian navigation techniques, represents exploration at its best. Conversely, European narratives of discovery often demonstrate how curiosity without humility or respect can lead to devastation. Reflecting on these historical examples can help guide us toward more thoughtful, ethical approaches in exploring unknown territories—whether on Earth or in space—today.

Manifest Destinies: Ethical Legacies in Contemporary Space Exploration

During his second inauguration, U.S. President Donald Trump declared, "We will pursue our Manifest Destiny into the stars," promising that American astronauts would plant their country's flag on Mars.[24] This invocation of Manifest Destiny—a term coined in 1845 to justify U.S. westward expansion through claims of divine sanction—illustrates how historical ideas still shape contemporary visions of space exploration.

Reflecting on the ethical contrasts between Polynesian navigators and European explorers reminds us that exploration never happens in isolation—it's intertwined with cultural beliefs, often influenced by religious traditions. To fully grasp contemporary ethical implications, it's essential to recognize exploration's often unnoticed relationship with pilgrimage. Both pilgrimage and exploration entail journeys into unknown realms associated with awe-inspiring power, whether divine or natural, prompting humility and ethical reflection. Acknowledging this historical connection helps us thoughtfully approach today's ethical questions about exploring unknown frontiers, especially in space.

Historically, pilgrimage referred to travel undertaken for religious or spiritual purposes. In medieval Europe, pilgrimage was one of the few accepted reasons for recreational travel.[25] Pilgrims set off to sacred sites such as Jerusalem, Rome, or Santiago de Compostela in pursuit of spiritual transformation, humility, and deeper connections to the divine. These journeys weren't merely about encountering new places but also experiencing inner transformation prompted by sacred landscapes.

The idea of a morally desirable pilgrimage continued beyond the medieval period. After the Protestant Reformation, religious freedom-seekers adopted the name "pilgrim." Instead of journeying east, they traveled west, aiming to establish a "New Jerusalem." They expanded into territories unknown to Europeans, believing that their spiritual mission justified hardships and moral compromises.[26] Like medieval pilgrims traveling to sacred sites, these settlers often viewed their relocation as fulfilling divine will, but they frequently overlooked ethical consequences.

Early scholarly analyses by Albert K. Weinberg framed Manifest Destiny as nationalist expansion rooted in perceived divine right.[27] More recent scholars, notably Anders Stephanson, emphasize how the doctrine deliberately combined religious beliefs and imperial ambitions during America's westward expansion.[28] Even as U.S. territorial expansion continued beyond continental boundaries into Hawaiʻi, Alaska, and various Pacific and Caribbean islands, the American public remained drawn to romanticized frontier visions. Religious studies scholar Catherine Newell argues compellingly in *Destined for the Stars: Faith, the Future, and America's Final Frontier* that religious values and visual culture motivating westward expansion also shaped popular visions of space exploration.[29]

Newell explains how Chesley Bonestell, whose artwork inspired generations of space enthusiasts, drew inspiration from earlier artists like Thomas Moran and the Hudson River School painters. Moran's depictions of the American West as an idealized "new Israel" profoundly influenced Bonestell's portrayals of outer space as humanity's divinely ordained frontier, subtly transmitting Manifest Destiny's ideals to twentieth-century audiences. According to Newell, Bonestell's paintings fueled enthusiasm for space exploration using techniques rooted in America's problematic history of westward conquest.[30] I've observed a similar religious motivation among some interviewees, who interpreted the U.S. reaching the Moon first as confirmation of divine favor.

Yet this distinctly American perspective of divinely ordained exploration is not universal among spacefaring nations. Japan, for example, approaches space exploration differently, shaped by religious traditions such as Buddhism and Shinto. Japanese narratives typically emphasize scientific inquiry, international cooperation, and harmony with the environment rather than conquest or territorial claims.[31] Recognizing these diverse cultural attitudes matters because America's particular legacy of Manifest Destiny, entwined with colonial practices of displacement, exploitation, and environmental degradation, raises significant ethical concerns for space exploration. Potential pitfalls include irresponsible resource extraction, disregarding the autonomy and rights of future stakeholders, and territorial or economic claims that might trigger conflicts over space resources or governance. Planetary protection advocates argue these ethical issues must be proactively addressed to avoid perpetuating detrimental colonial legacies in space.[32]

Today, private companies increasingly drive space exploration, introducing economic incentives reminiscent of historical colonialism. This situation prompts urgent ethical debates about mining asteroids, the Moon, and other celestial bodies for profit. Revisiting the awe astronauts experience when viewing Earth from space might serve as a powerful reminder of ethical responsibilities toward environments we explore. Astrophysicist Erika Nesvold emphasizes the importance of proactively confronting these ethical questions before crises emerge. As she states, "It may be too late to ask, 'Should we settle space?' but perhaps that's not the most important question. Instead, we need to ask, 'How should we settle space?'"[33] Nesvold's insights offer a crucial framework for humanity's responsible expansion into space—informed by ethical reflection on historical experiences with exploration and colonization.

The Ethics of Space Exploration

As a NASA-affiliated astrophysicist, Nesvold provides an insightful perspective on the ethical complexities facing contemporary space exploration. While working with governmental and private space organizations, Nesvold realized that clear ethical guidelines were rarely articulated. Instead, the prevailing attitude assumed that moral issues would naturally resolve themselves. Nesvold highlights the persistence of historical colonial attitudes—first developed during the Age of Exploration and carried forward through Manifest Destiny—as troubling influences on the modern space community and argues humanity has "an obligation to improve as a society before we spread beyond our home planet."[34]

While some envision human settlements on other planets, celestial bodies, or artificial habitats as a fresh start—a chance to create a more equitable society—others caution that Earth's current problems like racism, sexism, war, inequality, and corporate dominance are likely to persist wherever humans travel.[35] Both perspectives agree, however, that understanding our historical ethical failures is essential for preventing their recurrence in space.[36] Decisions made today about space exploration, colonization, and resource use will have profound and lasting impacts. Scholars warn against perpetuating Earth's flawed economic systems, particularly exploitative capitalism, into space, which could lead to environmental loss, resource depletion, and conflicts over celestial resources.[37]

Resource extraction in space, such as mining for water, rare metals, and Helium-3 on the Moon and near-Earth asteroids, presents significant ethical challenges. Optimistic views predict abundant opportunities, but more cautious voices note these resources are likely scarce, difficult to access, and their extraction could cause extensive environmental damage.[38] Many scholars also emphasize the necessity of diverse perspectives when making ethical decisions about space exploration.[39] Historians, philosophers, social scientists, Indigenous communities, and the broader public must all contribute to the discussion to avoid the narrow interests that have historically caused damage.[40] The rapid growth of space law underscores the urgent need for regulation and oversight; clear governance structures with enforceable expectations of behavior are critical for ensuring space remains livable and accessible for future generations.[41]

Privatization of space exploration further complicates the ethical landscape. The increasing involvement of commercial enterprises, motivated primarily by profit, differs substantially from the scientific goals traditionally emphasized by government agencies like NASA and Roscosmos.[42] Although humanity is unlikely to encounter Indigenous populations on nearby celestial bodies, there remain valid concerns about ecological damage, disruption of extraterrestrial microbial life, and cultural impacts—such as the potential loss or disrespect of heritage sites on the Moon or concerns about harm coming to the Moon itself, which holds spiritual significance for diverse Earth communities.[43] Learning from historical mistakes in exploration is imperative. Though private companies might resist oversight due to commercial interests, history clearly illustrates that unregulated and unethical exploration has long-term destructive consequences.

Exploration and the Future

Throughout this chapter, we've considered how exploration might best be defined and explored the ethical nuances that accompany it. While experiences of awe can inspire humility—fostering openness to new perspectives and guiding responsible exploration—history clearly demonstrates the dangers of losing sight of humility and respect for the unknown. The ultraview effect vividly illustrates how awe-inspired humility and ethically informed curiosity can positively influence humanity's approach to exploration, promoting responsible stewardship of the cosmos.

Astronauts who have experienced the ultraview effect—an intense, unsettling experience of awe and disorientation triggered by gazing outward toward the vastness of the Milky Way—often describe emerging with a profound sense of humility and uncertainty, reconsidering what they previously thought they knew about the universe. Unlike the overview effect, which inspires protectiveness, emotional connection, and deep affection for Earth, the ultraview effect emphasizes humanity's smallness, vulnerability, and lack of knowledge in the face of cosmic immensity. Both experiences, however, underscore the importance of humility and ethical responsibility in exploration, reminding us to carefully approach unknown frontiers with deep respect.

Moving forward, a thoughtful awareness of the ethical challenges addressed throughout this chapter can help ensure future explorations become genuinely inclusive, considered, and beneficial for all of humanity. Remembering lessons from history can further underscore the imperative of including diverse perspectives and maintaining careful attention to ethical considerations.

In our concluding chapter, we'll explore human endeavors—both in space and on Earth—that have shaped our pursuit of knowledge, emphasizing the importance of teamwork and cooperation in those efforts. We'll examine insights gained from exploring Earth's extremes—from the parched expanses of the Sahel to the astonishing depths of the oceans—and even the intricate mysteries of the human body. In doing so, keep in mind how awe, a realistic recognition of what humankind knows and does not know (as Socrates urged), and the willingness to collaborate and learn from others, can together guide us toward meaningful, ethical, and enlightening discoveries.

7 *Stepping Together into Uncertainty*

The Power of Collaborative Discovery

Throughout this book, I describe how many explorers and scientists move through the cycle of awe-humility-curiosity. Exploration without awareness of what remains unknown or the willingness to learn can lead to disaster. When researchers recognize their limitations and seek others with different skills, however, gaps can be filled, making projects more likely to succeed. I always love examples of this, particularly when a relieved astronaut finally arrives at their new "home" in orbit after a long spaceflight. When the first, uncrewed SpaceX flight docked with the ISS—an important step in solidifying its partnership with NASA—astronaut Anne McClain spoke about how collaboration and teamwork create unity. She said, "We are more alike than different. We can be united by a cause that is not based on fear, threat, or a common enemy, but rather on a bold endeavor—an insatiable curiosity to go beyond what is known and to do what has never been done. We humans were built for exploration, and we were built to do it together."[1] McClain is right: human curiosity inspires exploration, and exploration succeeds best when we recognize we need each other.

This emphasis on exploration as cooperative sets it apart somewhat from awe and humility, which often occur in solitude. The retired astronaut I call Theo has sought rich, wonder-filled experiences, on Earth and in space. When we spoke during his work on a project in California, he told me, "I climb the mountain out here every single night and my God, that's good stuff. I do better up the mountain than I do here in the office. Maybe that shouldn't be true, but it is. I'm addicted to climbing the mountain and I'm alone. There's no one there. The sunset is gorgeous, and the creatures are all there and there's no humans and it's a pretty nice addiction." Theo explained that the solitude, height, and view gave him a combined spiritual and mental lift beneficial to his work. It sounded like a great nightcap.

Conversely, another astronaut, one I will refer to as "Jackie," had an experience of awe that was shared with others but incredibly powerful all the same. Years ago, Jackie (who was sometimes stationed overseas) and some colleagues were undertaking some training flights during a total solar eclipse. There were two planes, each with a pilot flying in the front and a weapons system officer in the back. The flight team knew the eclipse was occurring later that day and routed their return flight so that they would be in the path of totality (the seconds or minutes when the Moon completely blocks the sun). Jackie felt incredibly lucky to be flying at that moment, to be in the air at the time of the eclipse, and to be in the right spot geographically. She and her team managed to fly into the shadow of the Moon as it blocked the light of the sun, and stayed with it for several minutes. While still in flight, they were able to take off their protective glasses and look directly at the eclipse, observing the undulating rays of sunlight that reached past the eclipsed disk like a fiery halo.

I have experienced total solar eclipses from Earth and been in total awe, captivated by the overwhelming "wrongness" of day suddenly becoming night (the horizon looks like sunset in all directions), and the eerie look of the Sun blacked out completely except for the waving tendrils of the corona (Latin for "crown"). There is emerging scientific evidence that these amazing cosmic sights evoke awe in ways that are measurable in the brain.[2] What Jackie experienced seems to have been even more intense. She said, "You could just look around and see light everywhere. You weren't so high that you could see shadow-lit ground, necessarily. But you could definitely tell blackness and light all around. I couldn't even imagine what it was going to be like. When you're looking at it, it's not static. The corona, right? You . . . could see the difference in it. It's alive."

Jackie described the solar eclipse from her vantage in a training jet as "Crazy cool. That's probably the coolest thing I've ever seen." Although at the time Jackie had not yet flown in space, she said she had talked with a member of her team with spaceflight experience. This person said that she considered flying in the shadow of a total eclipse to be on par with the overview effect in its ability to inspire awe. Part of what made the eclipse so impactful for Jackie was that, unlike Theo's experience on the mountain, she was with companions. They communicated through their headsets, marveling together at the strange sight. Sharing the experience enhanced it. This collaborative awe not only enhances emotional resonance but also fosters a deeper recognition of our own limits. Becoming aware of the gaps in our understanding encourages in us an openness and willingness to learn from others. Whether experienced individually or shared collectively, this recognition becomes vital when exploration moves beyond the familiar, requiring teamwork and cooperation.

Finding Our Place in the Universe

Humility, like awe, can be experienced individually or within a group. The Vatican astronomer I call "Carlo" had accomplished much but remained modest. He described his achievements as limited compared to those of his peers from graduate school and his fellow Jesuits at the Vatican Observatory with advanced degrees in subjects unfamiliar to him. Carlo also viewed humanity modestly, not seeing our species as especially accomplished. When we discussed whether God wanted humans to go into space, he described God as an adult and humanity as mere children, saying that space travel was "an important step in making us more responsible for our planet and thinking about the responsibility and care of other planets. The picture that comes to mind is a little kid who, when you want to teach them responsibility, you give them a puppy."

Astronauts who've experienced the overview effect—viewing our planet from space and realizing its fragility—often gain profound perspective about Earth and humanity's place in it. Many describe a clearer awareness of both their individual limitations and humanity's collective vulnerability. While astronauts frequently praise Earth's beauty, none argue that it's too large or too strong to need protection. The overview effect clarifies humanity's place in the universe, naturally encouraging collaboration—the recognition that protecting our world is something we must do together.

The ultraview effect similarly inspires humility and collaboration. Looking into space, astronauts, dark-adapted and overwhelmed by endless stars, recognize Earth's tiny scale compared to its surroundings. Theo argued that billions of "friendly planets" must exist, especially given life's adaptability. He explained, matter-of-factly, "Life is massively powerful. It lives in so many conditions. You look at the four million species and more we've got (on Earth). What can life do on billions and billions of stars? And so, yeah, there are billions and billions of planets out there that have life." Having seen the Milky Way as an "unbelievably solid" wall, Theo realized that life on Earth, seemingly boundless to us, is minuscule compared to the cosmos. This humbling awareness motivates us to collaborate, pooling knowledge and resources to explore deeply and effectively.

Why Exploration Thrives with Teamwork

While awe can be solitary or shared, and humility can be experienced individually or on behalf of our species, exploration is different. For exploration to truly have an impact, it must be a shared activity. Think how many times you come across something amazing and realize you really want to show it to someone else—a cool rock, a new restaurant, a weird video.

Let me give you another example. I have conducted fieldwork in different places around the world, usually as a solo researcher. I've often had the experience of disembarking an airplane or a train, trying to navigate the local language, and traveling on my own to the location where I will be doing research and introducing myself (it's exhausting, in case you're wondering). In many ways, the ethnographic work of an anthropologist seems like a solitary activity. In reality, though, I have always had support from others—a larger team who helped with things like funding, understanding the data I was collecting, advising me when I got stuck, and much more.

Some of my most interesting experiences have been when I have brought students into the field and they have contributed their ideas and analysis, often shedding new light on the research questions. Once when I was interviewing a shopkeeper in Rocamadour my student noticed a book behind the counter and brought it to my attention. Our ensuing conversation about the book and my interviewee's understanding of Rocamadour, particularly her ideas of the shrine's mystical origins, was enlightening. I would never have noticed it without my student's nudge.

Shared fieldwork is wonderful, but there are other types of teamwork in anthropological research. Increasing numbers of ethnographers are beginning to see the people we interview in our research as collaborators and contributors rather than subjects. Even publishing is a type of collaboration. When research is written up into books, chapters, journal articles, or conference papers, it becomes part of a collective body of knowledge other anthropologists can use. This means that even the "solo" exploration I have done on my own into religion and culture has depended on the cooperation and contributions of many others. My work becomes part of a group effort among anthropologists and other researchers to understand humanity.

Lessons from Astronauts: Teamwork beyond Earth

Astronauts illustrate very clearly how exploration depends fundamentally on openness, teamwork, and shared purpose—qualities that are indispensable not only in space but also in broader human endeavors. Their experiences emphasize the importance of defining a shared mission and establishing a collaborative mindset, especially when exploring challenging and unfamiliar environments. Every astronaut I've spoken to, an amazing collection of space explorers, has mentioned the necessity of teamwork and crew compatibility to accomplish an objective. Astronaut Jackie, whose solar-eclipse flight I mentioned above, had played sports in her youth and took her sense of teamwork in exploration from what she had experienced then. She told me, "If you think about your life, you want to contribute to the team, whatever that team may be. You know, a sports team or a family or the military or NASA or Earth. To be part of this effort that's going, and exploring, and trying to answer questions." Cara was another astronaut who believed that some kind of membership, whether in her church or in NASA, "helped me be a more calm person and maybe have a better mental health in my life, because I feel like I'm part of a team. I'm part of a group. I'm part of something bigger than myself."

Alan, a shuttle-era astronaut and big sports fan, talked about teams in two ways. First, he said, a space crew was like a team about to play a championship game. He explained, his voice getting louder, faster, and more excited as he continued:

> As you're about to go out the airlock, it's like when you're . . . an athlete about to take the field for a championship for the very first time, and you know two things: One, you might never get this opportunity again. . . . This might be your one and only chance to get that championship. And two, you know that whatever happens in the next couple of hours is going to define you either in a very, very good way or a very, very bad way, depending on how it goes.

For Alan, the act of exploration meant you owed something to your teammates and that your identity would forever be associated with your team. Teamwork involves recognizing and honoring our mutual dependence on one another.

Alan also talked about the cooperative elements of teamwork in a discussion of how one space station crew prepares another to take over the ISS (I honestly think about this example a lot because it seems so *normal*). He stated:

> The training does a really good job preparing you for almost everything, but there's some mundane things that it doesn't, like where exactly do we put different kinds of trash, little tricks for preparing the food so it doesn't make a mess and where we keep the wrenches when we're not using them. It's like if you're going to be a house guest for a while in somebody else's home, you've got to learn where . . . they keep the towels. You know, what if I run out of clean ones, where do I get more? . . . It's not tips on how to run the emergency procedures; it's more of the day-to-day life kind of stuff.

Alan made it clear that there was teamwork in high-pressure situations like space walks but also in the more mundane activity of welcoming a new crew to the ISS.

When I interviewed the astronaut I call Beverly, we agreed to meet in a coworking office in a southern coastal town. With the long drive and parking woes, I was a little late and Beverly got there before I did. I sent her a text from the parking lot to apologize. Instead of being irritated, she asked how I took my coffee—when I arrived, I was greeted with a latte and led to a quiet room where Beverly chatted with me like we were old friends. Personable, calm, and kind, it was clear that Beverly herself must have been an excellent team member during her space missions and beyond.

Beverly had been thinking about what it meant to be part of a team or, more specifically, a crew, for a long time. She was impressed by the fact that the ISS was a shared international effort and felt strongly that it could be a model for similar cooperation on Earth. After all, what is our planet if not "spaceship Earth"? That said, there were some challenges to getting our population to behave more like the crew of an exploratory mission. Beverly mused:

> I think one of the things that's missing down here on the planet itself is that collectively we haven't set our mission. We haven't established our mission, this "greater good" mission that includes everyone. We haven't done that yet. Whereas when we build a space station, even when it's with fourteen other countries, we come together first and we say, "Okay here's the mission in the space station." We've got a crew code of conduct . . . that includes the ground, how we're going to communicate and . . . the rules of engagement. We don't do that here . . . how do we use this model of what we've done in space? We've built the most complex thing we've ever done in space. It's not like we did it in our garage, it's in space. And for the last twenty years, fifteen countries have peacefully, successfully, quietly done this together.

I find Beverly's argument here very inspiring. It's easy to forget sometimes that the ISS, whose replacement apparently won't be a NASA project,[3] was built and maintained through international cooperation. Beverly strongly believed it can serve as an excellent model of teamwork, particularly if more nations were included to establish a shared goal. She admitted that she didn't think "save the planet" was a strong enough mission to motivate most earthlings, but maybe "save *humanity*" would. She sighed, "The selfish motivation works so much better. It really needs to be 'save *us*.' Because the planet will survive us." Beverly explained that even something that could wipe out all humans couldn't actually harm the Earth. She considered, "The Sun could [even] take our *atmosphere* away, but the *planet* [will] still be here. *We* won't."

This recognition of humanity's collective vulnerability makes the necessity of collaboration abundantly clear—we need one another to survive and thrive, both here on our "spaceship Earth" and beyond it. Isolation and alienation, whether personal or political, are deadly. The astronauts' experiences we've just explored highlight two essential points: successful exploration depends on recognizing our mutual dependence, and the perspective gained through space exploration can inform our approach to global challenges on Earth. The following examples reinforce and expand on these insights: rediscovering ancient navigation techniques through cross-cultural collaboration, uniting interdisciplinary teams in deep-sea exploration, leveraging Indigenous ecological knowledge to respond to climate change, and working together to meet the overwhelming challenge of preserving human health in outer space.

Rediscovering Wayfinding: Mau Piailug and the Hōkūle'a Voyages

Ancient Polynesian wayfinders settled vast regions of the Pacific through methods grounded in humility, careful observation, and openness to uncertainty. Many believed that the traditional navigational system had been lost to history. A modern revival of these techniques, however, illustrates how collaboration can rediscover and reanimate valuable cultural traditions, allowing contemporary explorers to conquer new challenges together.

In the 1970s, there was a resurgence of interest in Indigenous Hawaiian culture, part of which centered on ancient traditions of navigating the oceans. At the time, many scientific theories suggested that the initial settlement of the vast Pacific region known as Polynesia had been accidental, with settlers passively carried there by ocean currents rather than through active navigation. The oral traditions passed down to many young Hawaiians, however, contradicted this idea with their mentions of purposeful voyages by canoe.

According to the website of the Polynesian Voyaging Society (PVS), in the 1970s "cultural extinction felt dangerously close to many Hawaiians." In one attempt to reverse this decline, members of the PVS built a boat, the Hōkūle'a, using ancient techniques. That could have been the end of it, but they realized they were missing vital information. Willing to admit the ignorance of their members and move forward with open minds, "the Voyaging Society looked beyond Polynesia to find a traditional navigator to guide Hōkūle'a: Mau Piailug, a navigator from a small island called Satawal, in Micronesia. He agreed to come to Hawai'i and guide Hōkūle'a to Tahiti." The PVS website continues, "Without him, our voyaging would never have taken place. Mau was the only traditional navigator who was willing and able to reach beyond his culture to ours."[4]

How did Mau Piailug, whose given name was Pius Piailug, come to have this knowledge? His birthplace, Satawal, was in the western part of the Caroline Islands, which had been neither colonized by Western powers nor considered strategically valuable for nuclear testing. Simple luck left the island undisturbed, permitting this traditional knowledge to survive and be passed on to Mau as he grew up. Initially, his teachers used pieces of coral to represent stars for navigation, but eventually Mau became adept enough to memorize an old and carefully guarded map of Micronesia "that included the names and positions of stars, habitats of birds and fish, and routes through reefs and atolls."[5] His mastery, rooted in

FIGURE 8. Polynesian wayfinders navigating by stars and the movements of nene geese and the water, showing the connection between humility, observation, and exploration. Illustration by Karl Tate.

deference toward his elders and careful attention to detail, prepared him uniquely to revive a nearly lost tradition.

When Mau was approached by the Hawaiians, such cross-cultural sharing had never been done. Even some members of his family were against allowing non-Micronesians to learn their practices of traditional navigation. Opening his heart to the PVS members, however, Mau eventually agreed to teach five Hawaiians and eleven Micronesians the information and techniques known only to those with the title *pwo*, bestowing this prestigious honor on his students in the process.[6] Mau and his pupils ultimately traveled from Hawai'i to Tahiti using skills passed down to Pacific Islanders over centuries, skills once thought irretrievably lost. The success of their expedition revitalized interest in Indigenous culture in Hawai'i, leading directly to the creation of the Office of Hawaiian Affairs and the decision to make Hawaiian an official language of the state.[7]

In 1978, confident in their recent success, a group of PVS members attempted to navigate the Hōkūle'a to Tahiti. Mau wasn't included, and a few hours into the journey the boat capsized. Eddie Aikau, a revered lifeguard and surfer, set out for help with a surfboard but was never seen again. The rest of the sailors were rescued by the Coast Guard. This could have been the end, but it made Nainoa Thompson, one of Mau's most prominent students, resolute. He spent more time learning the traditional methods. Two years later, more practiced in the techniques Mau had taught him, Thompson and his crew successfully sailed the Hōkūle'a to Tahiti without compasses or other modern devices. They undertook the journey to correct the previous failure but also to honor the heroic Aikau. The success of this expedition was deeply emotional. Upon safely reaching Tahiti, Thompson reacted not with celebration but with quiet reverence. He later reflected: "I just went into a quiet, dark place and told Eddie we pulled it out of the sea. . . . There's no high fives. It's too profound."[8] Thompson's response, then, reflected sincere respect and quiet reverence, intertwined with the awe inspired by Mau's teachings and the solemn memory of Aikau's sacrifice.[9]

The modern revival of Polynesian wayfinding demonstrates how self-awareness, openness to cultural exchange, and respect for the knowledge of others can inspire meaningful exploration. Mau Piailug's willingness to share his cultural knowledge, combined with Nainoa Thompson's determination, preserved traditions nearly lost to history, reconnecting people across generations and cultures. While Polynesian navigators relied on approaches tied to direct human observation, other explorers have taken a different path, using cutting-edge technology to venture into environments where humans cannot survive unassisted. The same principles of humbleness and collaboration that reanimated wayfinding traditions can also guide technology-based explorations, as we'll discover in the groundbreaking (water-breaking?) journey of the Trieste bathyscaphe to the deepest point in Earth's oceans.

Into the Challenger Deep: Technology, Teamwork, and Trust

Just as Polynesian navigators relied on careful observation of the ocean's surface and the sky—methods tied to humility, attentiveness, and awe—marine biologist Rachel Carson found a similar sense of wonder by contemplating the vastness of the sea and humanity's place within it. In her 1951 book *The Sea around Us*, she writes that while at sea, a human being can understand the "true nature of his planet" when he "watches day after day, the receding rim of the horizon, ridged, and furrowed by waves; when at night he becomes aware of the earth's rotations as the stars pass overhead; or when, alone in this world of water and sky, he feels the loneliness of his earth in space."[10] This awareness of being on a planet is a tremendous source of awe, one that can be experienced where the sea appears to become one with the sky, linked together by an invisible horizon. Carson's descriptions recall the planetary awareness of the overview effect and the dazzling incomprehension of the ultraview effect. Yet beneath the ocean's surface, these powerful experiences of awe deepen, inspiring exploration into realms that challenge not only human perception but also technological ingenuity.

Awe can be evoked not only at the ocean's surface but also when attempting to understand its *depths*. Surprisingly, I have felt the awe-inspiring truth of the ocean's depth from the comfort of my couch. When Luke was twelve, he showed me a website that simulates a deep-sea dive and shows the depths attainable by different animals.[11] The polar bear, for example, is limited to fewer than 30 meters under the surface, while the killer whale can dive to more than 100 meters. As we descended, the image of the ocean grew darker and darker. We passed the lowest depth a scuba diver had ever reached, 332 meters, and continued. I began to feel uncomfortable as we (in our imaginations) sank even deeper and, to my growing bafflement, continued to find living things that could withstand an unimaginable amount of pressure in an alien landscape of weight and darkness.

We eventually reached the website's version of the Challenger Deep in the Mariana Trench, which is 10,891 meters, or 6.77 miles, below the ocean's surface. Along the way, Neil Agarwal, creator of the site, tells the story of the 1960 expedition of the Trieste bathyscaphe, a submersible craft similar to a submarine, which was piloted by explorers Jacques Piccard and Don Walsh. Crowded together in a perilous craft that could implode at any second, they reached the Challenger Deep after a descent of almost five hours.[12] If Jacques Piccard's name sounds familiar, that's because I mentioned his father, Auguste Piccard, in Chapter Two, when discussing the first successful human attempt to reach the stratosphere. One Piccard went high, the other went low, but both followed the pull to explore places of intense wonder.

Of course, the successful dive to the Challenger Deep wasn't just the work of two men. The U.S. Navy provided support and funding for the expedition, as well as personnel. The bathyscaphe was assembled at the Swiss Federal Institute of Technology, and much of its design was developed by Auguste. Walsh and Jacques Piccard also relied on a team of engineers and studies of the oceanographers and researchers who had mapped the ocean floor of the Mariana Trench. Finally, the ocean tug USS *Wandank* (ATA-204) and her crew acted as a support system for the explorers, pulling them from the sea after their slow ascent back to the surface.[13]

While vessels like the Trieste represent remarkable feats of engineering and exploration, recent events—particularly the tragic implosion of the Titan submersible in 2023[14]—have cast a shadow over similar deep-sea expeditions. Despite troubles, research conducted by these technological marvels has greatly expanded human knowledge, from previously unknown ocean species to the engineering required to survive extreme underwater environments.[15] Intriguingly, some of the same curious minds who have explored space have also ventured to the depths of the Challenger Deep, including retired NASA astronaut Kathleen Sullivan and

FIGURE 9. The Trieste bathyscaphe during its descent to the Challenger Deep, evidence of the depths of human curiosity and technological ingenuity. Illustration by Karl Tate.

Roscosmos spaceflight participant Richard Garriott de Cayeux.[16] Yet while the Trieste exemplified exploration driven almost exclusively by advanced instrumentation, our next example offers a fresh, contemporary approach, one that integrates advanced technology with traditional human observation and knowledge. This synthesis of the ancient and the modern, as we'll see in Hindou Oumarou Ibrahim's innovative climate mapping, offers a promising model for future exploration.

Mapping Climate Change: Indigenous Knowledge and Satellite Technology in Chad

Some divide the African continent into two sections: Saharan and sub-Saharan. The Sahel region, which includes parts of Chad, is the transition area between the deserts of northern Africa and the savannahs and forests to the south. Indigenous leadership and traditional ecological knowledge from the Sahel, in a project spearheaded by Hindou Oumarou Ibrahim, is directly influencing responses to climate change. Ibrahim's innovative work illustrates how traditional knowledge can not only inform local decision-making but also shape global scientific policy.

Ibrahim, a member of the Mbororo people of Chad, was born in 1984. She is the founder and coordinator of the Association of Peul Women and Autochthonous Peoples of Chad (AFPAT), a women's rights and environmental protection organization that she established when she was just fifteen years old.[17] In the early 2010s, through AFPAT, Ibrahim began to work on projects that focused on creating 3D maps of the Sahel that incorporated traditional knowledge from the people of that area, including "land use, traditional routes of cattle migration, ecosystem features, and biodiversity information."[18] As part of these projects, members of local populations study satellite images and contribute their own observations and memories, marking significant places and objects. This is a community-wide undertaking and includes people from multiple villages of different ages and genders. The projects make the importance of different areas of these maps clear to the wider groups, promoting not only the sustainable use of resources but also peaceful coexistence among the communities involved.[19]

Much of the success of these projects comes from Ibrahim's own philosophy. She grew up in a pastoralist society, in which cattle are kept along with sheep and goats. Older women among the Mbororo were able to analyze nature's signals, such as the behavior of birds or insects or the movement of clouds, to predict rain or determine when to move herds or plant crops. While using a Geographic Information System (GIS) and other techniques to study the climate, Ibrahim realized they could be used in tandem with the learning passed down in her own community.[20] Much like the Hawaiians who sought out traditional wayfinding knowledge, Ibrahim learned traditional methods. In her case, however, it was to combine them with satellite-based techniques. In both examples, the older systems were not underestimated or ignored but respectfully maintained by people who understood that newer processes, while impressive, cannot fully replace or substitute for approaches worked out over centuries.

Ibrahim's gift for local activism has taken her far beyond Chad; she has even briefed the United Nations Security Council on climate change multiple times, using information collected from the local people in the Sahel to inform international initiatives. Global recognition and multiple awards have allowed her to expand her work to the neighboring country of Niger, also in the Sahel, providing information and inspiration to communities deeply affected by climate change and giving them a chance to thrive.[21] Ibrahim is modest when addressing a worldwide audience, making it clear that the information she shares comes from her community rather than herself. In her speeches, she talks about what each group in the Sahel brings to the table—women who know about drought-resistant plants or fishermen who understand the rhythms of local bodies of water—creating solutions enacted by the Indigenous community instead of being imposed from the outside.[22]

The high value placed on traditional approaches to exploration in Polynesia and Chad doesn't mean that expeditions undertaken with advanced technology or to environments inaccessible by conventional methods can't also benefit from collaboration and openness. Even in highly technical contexts, successful exploration depends on recognizing the limits of individual knowledge and the importance of working together.

Keeping Humans Healthy in Space: An Exercise in Collaboration

The reality of climate change, once understood, can be overwhelming, evoking awe that is disturbing rather than pleasurable. This overwhelm sometimes results in paralysis and inaction: "I don't understand and I can't do anything anyway, so I'm going to ignore it." Yet for others, like Hindou Oumarou Ibrahim, such challenges inspire action. She recognized her knowledge gaps and embraced diverse sources, combining traditional Sahel wisdom with satellite technology. While climate change directly affects human survival on Earth, space exploration faces parallel challenges, requiring survival and innovation in extreme environments. Both fields benefit from interdisciplinary collaboration and mutual insights.

One of my favorite research venues is NASA's Human Research Program Investigators Workshop (IWS), an annual meeting for experts tackling human adaptation to space environments. The IWS brings together psychologists, doctors, engineers, and even anthropologists, highlighting spaceflight's incompatibility with human biology. Space travel triggers muscle and bone atrophy, fluid shifts, sensory disorientation, and serious conditions like SANS. Astronauts face radiation, sensory deprivation, and disrupted sleep, relying on medical testing and medications to manage severe health challenges.

Given these difficulties, why pursue human spaceflight at all? IWS attendees emphasize that exploration, including of space, reflects a fundamental human impulse. "Solomon," a senior administrator, explained, "People will always go in a direction where people haven't been before. . . . I think it's important because it's inevitable. It's not like something we will choose." "Badra," a medical student and aspiring astronaut, added, "Exploration of any kind is important just to broaden your horizons and see what's beyond what you already know."

Others emphasized the practical benefits that space exploration holds for Earth. "Joanna," an expert in critical care medicine, noted, "What we've learned about ourselves will hopefully help us manage disease and potentially extend life . . . and understand that Earth is fragile." Quinn, whom we met previously, saw human spaceflight as essential for mastering complicated problems in general and useful for addressing broader issues. He explained, "Complex human spaceflight [has a] large number of highly interconnected factors [making it] a laboratory for the study and implementation of complex systems." Nick, a medical engineer studying circulatory systems in microgravity, affirmed this view: "Everything we learn about trying to survive and thrive in space can be directly applied to everybody else back on the planet."

Ultimately, keeping humans safe in space—on the Moon, Mars, or beyond—is a test of human limits and ingenuity. Each obstacle, from radiation exposure to psychological isolation, requires cooperation across disciplines to overcome. Those working to ensure human safety in space are pursuing exploration at its finest: taking a courageous, collective step into the unknown, driven by a recognition of mutual dependence.

Beyond Manifest Destiny: A New Ethic of Exploration

Throughout this book, we've explored how awe, humility, and collaboration enhance exploration. When explorers acknowledge what they don't yet know, they create conditions conducive to cooperation and mutual learning. We've seen this in the examples of Polynesian wayfinding, the technological daring of the Trieste bathyscaphe, and Hindou Oumarou Ibrahim's synthesis of Indigenous wisdom and satellite imagery. Each shows that exploration, while possible individually, becomes richer, more insightful, and ethically grounded through teamwork and openness.

These examples offer more than inspiring stories—they represent a meaningful alternative to harmful legacies such as colonial conquest and exploitation, and destructive ideas like Manifest Destiny, notions that still linger in contemporary politics. The ultraview effect, with its humbling view of humanity's place within the cosmos, provides a model of exploration marked by mutual respect, genuine curiosity, and profound openness to learning from one another and from our universe.

I encountered a striking example of this perspective at the Mojave Air and Space Port. A test pilot I call "Orville" described his creativity as rooted in an intergenerational chain of curiosity. His outlook was shaped by mentors and heroes across many fields. He admired his groundbreaking employer, whose inspirations came from twentieth-century rocket designers influenced by filmmakers like Fritz Lang and visionaries like Jules Verne. Orville reminded me that exploration is fundamentally interconnected, each discovery building on the work of those who came before.

While Orville's influences stretched back in time, the ISS exemplifies contemporary global cooperation. Since 1998, the ISS has combined knowledge, resources, and ingenuity from more than a dozen countries. Individual nations have built their own space stations, such as the Soviet Union's, later Russian Federation's, Mir or China's Tiangong, showing that collaboration is not required. Yet the ISS offers richer scientific opportunities, through shared resources and costs; technological resilience; and diverse, multinational insights. The willingness of participating countries to acknowledge their own knowledge limitations has transformed the ISS into a powerful symbol of collaboration.

Astronauts frequently describe how seeing Earth from space evokes awe and vulnerability, encouraging exploration that respects the unknown and values human interdependence. This perspective becomes increasingly important as humanity faces urgent global challenges like climate change. Medical researcher Nick emphasized that solving complex problems in space medicine—such as understanding microgravity and radiation impacts—benefits significantly from interdisciplinary collaboration and humility. These qualities are equally crucial for addressing climate change, underscoring how Earth-based and space-based explorations are deeply interconnected.

While experiencing awe in the presence of the cosmos, it's worth appreciating one's capacity to reflect on these experiences. A space researcher I call "Ken," an eccentric engineer who has explored both the highest heights and deepest depths, demonstrated a form of awe unique among my interviewees. Ken was in awe of humanity itself, viewing us as a miraculous, profound, space-faring lifeform. Ken's perspective reminds us to equally value the extraordinary fact of human existence and consciousness. As Ken put it: "There is a universe beyond the Earth, and humans have an unbelievable connection to the universe. In almost every culture, people have dreamed about traveling beyond the Earth for millennia, documented for centuries at least. The emergence of this species into the universe as a multi-planet species, our drive to understand how this all came to be, is completely remarkable."

Give yourself moments of awe, feel humble and curious, explore widely and deeply, and never lose sight of your own remarkable, improbable existence—as much a part of the universe as its endless stars.

Opening Vistas

Conclusions and Suggestions

At the end of a 2006 *New York Times* opinion piece, physicist Brian Greene wrote, "Exploring the unknown requires tolerating uncertainty."[1] Greene was explaining why attempts to describe the universe—like string theory—remained works in progress and constantly changed with new data, but his concluding remark beautifully encapsulates what I've found through the research that went into this book. My time spent getting to know astronauts and others showed me how important the experience of awe can be for appreciating the universe we live in and overcoming a tendency to live as though we already know everything there is to know.

Awe is crucial, but it is only a first step in a path that leads from wonder to humility, and from humility to curiosity-driven exploration. Our species is constantly facing new dangers and challenges that may threaten our survival. It is my hope that stopping every so often to engage with things that are far larger than our planet can help us recognize our limitations and build on them in a way that allows our knowledge to grow.

This book mirrors a pattern I've seen in many members of the space community, a cultural group that spends more time than average contemplating the reality of our tiny place in the universe. Awe transforms human beings, shaking us up so that our perspectives are clear and ready for new information. Flight, especially spaceflight, is capable of producing awe in humans for multiple reasons, not the least of which is how foreign flight is to our DNA and the experiences of our long-ago ancestors. Humans, unlike birds, bats, and insects, have no wings, but our inventions let us fly anyway—and it is thrilling. These thrills, however, are punctuated with moments of wonder or strangeness, such as the ultraview effect, when we realize the extent of the vast mysteries we are not equipped to understand.

One of the side effects of an awe-inspiring experience is humility. While not everyone is comfortable with humility, the wisest among us recognize our limits both as individuals and, as astronauts have seen firsthand, inhabitants of a small planet that functions as the only life-support system we have in a bewilderingly large universe. Traditional cultures and the spiritual sages from a variety of religions have frequently emphasized the importance of being humble, especially for fostering personal and collective growth.

Humility makes us recognize how much more there is for us to know, and when this feeling is embraced, it can serve as a catalyst for exploration. True exploration, whether on land, in the seas, in the sky, or beyond, must be undertaken with an open mind and an awareness of ethical pitfalls that are sometimes easy to ignore. It also requires teamwork—the support of the group—in a way that allows different experiences and knowledge to be woven together into something new and powerful.

Awe, humility, and exploration can be experienced on their own or work together in the kind of cycle I've seen at play in astronauts like Theo, Beverly, and Zack. Below, I have gathered some suggestions to encourage you to experience awe, nurture a sense of humility, and welcome adventure in your own life. These lists are not exhaustive, and perhaps many of them are already important to you. Still, I hope they will inspire you to reap the benefits of the ultraview effect right here on Earth.

Seeking Awe

LOOK AT THE SKY. Even if you live in an area with a lot of light pollution, there are ways to experience the amazement of stargazing. Many communities have observatories with public viewing nights, and a planetarium can provide awe-inspiring views year-round. If you can, travel away from city lights or find a dark sky park, especially when there are annual meteor showers taking place, like the Perseids in mid-August, the Leonids in mid-November, or the Geminids in mid-December. Telescopes and even binoculars can bring everything into focus.

SPEND TIME IN NATURE. Life itself is incredibly mysterious and can be a potent source of awe. Listen to birdsong; follow animal tracks; listen to the wind rustling the leaves; or contemplate the sheer numbers of living things that exist underground, in the water, and even in the clouds. Even places that seem deserted teem with life, and surprises await if you're patient.

LOOK WITHIN. You are a vibrant ecosystem. Think about the lifeforms that make you their home, from harmless eyelash mites to the gut microbes that help you digest an apple. The cells that make up your body are also living things, living, reproducing, and dying. Check yourself out with a microscope or simply take some time to think about the complex ecosystem within you.

FIND RELIGIOUS OR SPIRITUAL AWE. If you're a believer, seek out sacred spaces that are naturally occurring or have been built by humans. Listen to moving pieces of music or immerse yourself in visual art, like paintings and sculpture. Look at amazing NASA images like the Hubble Deep Field photo. Read religious scripture or accounts of other people's spiritual experiences.

FIND WHAT EVOKES AWE IN YOU. Experiment with different experiences and information until you find something that blows you away. Flight surgeon Quinn was amazed by the mysterious world of quantum physics. In addition to the vastness of the universe, I tend to be awed by microbiology and evidence of prehistoric human societies, like ancient cave paintings. Go to museums, read books, and watch documentaries until you know what excites and amazes you.

Cultivating Humility

ASSESS YOURSELF. As objectively as possible, figure out your skills, your knowledge, and any areas where you could be considered an expert. Write a list of what you still don't know in these areas and who knows more than you do. Try to adopt the "Beginner's Mind" perspective from Chapter Five as you recognize where your knowledge still has room to grow.

VALUE DIVERSE KNOWLEDGE. Reflect on the kinds of insights you value. Are you more convinced by knowledge that comes from scientific experiments, field studies, or from people's firsthand personal experience? Are you more swayed by the wisdom of the old or the fresh outlook of the young? Think about what can be learned from each perspective and the varieties that exist in each. If personal accounts are more convincing to you than scientific findings, think about why that is and listen more to what researchers have to say. If you are less convinced by personal stories, try to see them as a source of data that is more deep than broad. Be open to knowledge from a wide array of disciplines and from people whose perspectives and life experiences are very different from your own, especially cultural groups about which you know little.

BE WILLING TO CHANGE YOUR MIND. Humility requires constant questioning and reevaluation of what is known. We live in a time when political, religious, social, and other kinds of identity often set limits on what it's okay to think or believe. A humble perspective means that your beliefs or understandings are never set in stone. New information can be considered, and new conclusions can be reached.

READ WIDELY. Reading is one of the best ways to take in new information, and a willingness to take in new information is itself an admission of humility. Read for pleasure and information, but also make an effort to read things that intimidate you. Make a particular effort to read about things you devalue, or think aren't worth exploring, even if it's just to learn why some people find a topic interesting. Be sure to mix up your genres, too. Read memoirs, poetry, textbooks, newspapers, legal decisions, children's stories, and scientific journals. So much is available online now that was never available before—you can even read the full Apollo mission transcripts, and they are fascinating.

ADMIT TO IGNORANCE. Practice saying, "I don't know about that," "Tell me more," "This is new to me," or "I've always wondered about that" in conversations. Acknowledge areas of limited human knowledge. No one really knows, for instance, why music exists; why people cry from emotion; how bicycles balance; what dark matter is; or the full diversity of animal, plant, and fungal species in our oceans and underground. Become more comfortable with your own lack of knowledge and the limits of human understanding.

Embracing Exploration

SEEK NEW PLACES. It may not be the most original suggestion, but I urge you to make short trips or longer voyages as often as you can. Whether you can occasionally get off the bus at an unfamiliar stop or have a rare chance to sail off to an unfamiliar continent, take those opportunities and embrace them. Going somewhere you've never been before will give you knowledge and perspective that will never leave you. It's also a great way to meet new friends.

DIVE INTO THE UNFAMILIAR. Exploration doesn't always require travel—it can begin with small, intentional steps into unfamiliar territory. Watch movies you wouldn't normally choose and listen to music you've never heard before. Try sports or hobbies that seem strange at first. It's even possible to experience the unfamiliar by changing your routines. Try commuting to work via a new route or join a club that meets weekly. Let curiosity about the unknown pull you into novel experiences.

VOLUNTEER. Volunteering creates an opportunity to leave our routines behind to enlist in a joint venture. Connecting with others to help reach a common goal allows you to make new discoveries and work with a team, whether you're volunteering for an event, an election, a soup kitchen, a nursing home, working with refugees, or rehoming animals. You gain a new form of expertise and a new community all at once.

ATTEND COMMUNITY EVENTS AND LECTURES. If you live somewhere with museums or bookstores that invite speakers, or if your community hosts cultural events, make a point of participating from time to time. When you go out into your world, you give yourself an opportunity to expand what you know and who you know. If you're more isolated, look for online talks and events, which have become more common. It's easier now than ever before to explore the world and connect to others using your phone or computer.

WORK ON VIEWING THE WORLD POSITIVELY. As a species, humans have a built-in tendency to view the world as a dangerous place. The same instincts that protected us from dangerous predators like big cats and wolves keep us on high alert as we go through our daily lives. Having both evolutionary and social reasons to be wary of the world makes a lot of us stick to the familiar and avoid adventure. While some fears and dangers are real, exploration and awe help us balance our caution with curiosity, allowing greater engagement with the world.

. • .

In practical, everyday terms, cultivating awe and humility equips people to solve problems more effectively, foster stronger relationships, and build healthier, more resilient communities. Awe reminds us that we do not (and cannot) know everything. Humility helps us stay curious, adaptable, and willing to learn from one another. Together, these qualities form the foundation of a wiser, more cooperative civilization, one capable of asking better questions and embracing a more expansive view of what it means to be human.

At a time when public and governmental support for science and exploration has been eroded, it is especially important to reaffirm why these pursuits matter—not as spectacle or escape, but as a profoundly human pursuit. Space exploration stretches the boundaries of understanding across science, philosophy, and emotion. It reveals the limits of our control, the depth of our ignorance, and the scale of our shared reality. Looking outward from space fosters insight that is not just intellectual but also emotional and existential—something only lived experience can fully reveal. Only human presence—embodied, reflective, imperfect—can carry meaning into the cosmos and bring perspective back home.

In an increasingly complicated world, awe, humility, and exploration not only enrich everyday experiences but also provide a path to deeper insight and transformative growth. These traits help individuals navigate routine interactions and build stronger communities, but they also foster the kind of expansive thinking so urgently needed in times when curiosity and scientific inquiry are devalued and sidelined. Awe interrupts familiar assumptions, illuminating the vastness that lies beyond ordinary horizons. Humility encourages grounding in reality, while inviting lifelong learning and openness to change. Exploration, propelled by curiosity and cooperation, transforms uncertainty from something daunting into something exhilarating—sparking fresh perspectives and bold new questions. Cultivating these traits together helps individuals not just face the future but also shape it—deepening connection, enriching experience, and supporting the growth of a wiser, more inclusive world.

Ultimately, the ultraview effect is not just a window into the cosmos—it is also a mirror. It reflects our smallness, our unfinished knowledge, and our potential to become something more. It invites those who are willing to look outward to become the kind of explorers this moment demands: curious, humble, and wide awake to the unknown.

Acknowledgments

I want to thank the many people whose help was invaluable while this book was coming together. At the top of the list I must place my husband Glen E. Swanson, space historian extraordinaire, whose enthusiasm, endless knowledge, and willingness to serve as principal "informant" somehow lured me into researching the space community in which he plays such an active part. When we met I thought it was *so cool* that you were a space historian and just look how everything turned out. This book would not exist without your willingness to teach me the customs of your people. I love you! I'm also grateful to our son, Luke Swanson, who loves history as much as his dad. Although you prefer ancient Babylonia to the Apollo missions, you've always been very patient as your parents dragged you from space centers to spaceports and made hanging out with astronauts a regular part of your childhood.

Thanks to Kate Marshall, the University of California Press editor who has been truly *ultra*. You provided the genesis of this idea and stayed with me all the way, despite my anthropological quirks. I'm also grateful to the UC Press Editorial Committee, who supported this book and facilitated its development. Thanks to Karl Tate for creating the gorgeous art that illustrates its concepts.

I also want to thank my colleagues and friends from the Grand Valley State University Anthropology Department who've put up with this weird take on our discipline. Thanks also to the interdisciplinary folks at Brooks College (whose approach has clearly rubbed off on me), to GVSU's Center for Scholarly and Creative Excellence, and to the people in the Dean's office in the College of Liberal Arts and Sciences and in the Provost's office, who helped me get my all-important 2019 sabbatical under way.

Plenty of colleagues outside of GVSU deserve thanks, including more people connected to the University of California, San Diego's Anthropology Department than I can name. Thanks especially to Kevin Birth, Julia Offen, the UCSD folks in a certain Facebook writing group, and particularly to Jon Bialecki, who told me that I basically *had* to do this research because of my unusual access. Gratitude to space-focused colleagues in the American Anthropological Association. And to the deep-from-childhood friends who weren't all that surprised to see me become a space anthropologist: Christine, Diane, Robin, and others (yes, you too). Big thanks are due to Sarah, Eric, Eileen, and other encouraging friends, to the whole *Roger That!* team, and even to my students, more willing than you'd think to indulge me in my space-related tangents.

Additional gratitude is due to the wonderful space community members I've gotten to know during the last few years. Your numbers, like the stars, are nigh uncountable, but I want to give special thanks to Niamh, Mindy, Yvette, Paul, Chris, Alice, Frank, Emily, Roger, Dwayne, and Margaret. I also want to thank the Explorers Club for their support, particularly the wonderful folks of the Chicago/Great Lakes chapter and Al and Bob, who generously sponsored my membership. Thanks to the *Specola*, APL, JPL, the HRP, JSC, KSC, and other groups and agencies who welcomed me so warmly.

A very special note of gratitude to the massively popular UK rock band Kasabian, whose song "T.U.E. (The Ultraview Effect)" served as a surprise soundtrack to this book (you can find it online). Thanks for researching space stuff during the Covid quarantine, Serge Pizzorno, and for aptly dubbing my ultraview effect "the Milky Way version" of the overview effect in a Radio X interview. It's an unexpected gift to have a rock song, especially an extremely cool and psychedelic one, named after a term I coined.

Above all, thanks to the rest of my family, including my father, Norman, and sister, Carol, both too recently lost; my brother, Carl (who remains steadfast); and my other treasured kinfolk, near and far. Finally, I wish to thank my late mother Carlyn. I drafted the structure of this book on the day you left us, just before I spent my last morning at your side. Your spirit infuses it, and your quest for truth made it possible.

Notes

Introduction

1. The "limb" of the Earth is the visible atmosphere as seen from space.

2. Howard, *Blast Off!*, 9.

3. For discussions on how leaders' reluctance to admit uncertainty or mistakes can limit growth and effective decision-making, see Fung, "Why Can't Politicians Ever Admit They've Been Wrong?"; Grant, *Think Again*; and Kahneman, *Thinking, Fast and Slow*.

4. Suebsaeng, Dickinson, and Perez, "Trump and Elon's 'Pointless Bloodbath'"; Higginbotham, *Challenger*.

5. Cunningham, "New Disneyland Ride Previewed," 17.

6. Suckow, "NACA Overview."

7. Meyerowitz, *Not June Cleaver*.

8. Weitekamp, *Right Stuff, Wrong Sex*.

9. In a happy turn of events that was also good publicity for Blue Origin, a private space company, both Dwight and original Mercury 13 participant Wally Funk were finally included in commercial spaceflights, Funk in 2021 and Dwight in 2024.

10. Paul and Moss, *We Could Not Fail*.

Chapter 1. The Essence of Awe

1. Powys, *Llewelyn Powys*, 82.

2. Merriam-Webster, "Awe," https://www.merriam-webster.com/dictionary/awe.

3. Nakayama et al., "Individual and Cultural Differences in Predispositions to Feel Positive and Negative Aspects of Awe," 773.

4. Carrive, "Le sublime dans l'esthétique de Kant," 72.

5. Keltner, *Awe,* 8.

6. Whitman, *Leaves of Grass*, 221.

7. Watterson, *The Days Are Just Packed,* 101.

8. Clark, *Einstein*, 755.

9. Sagan, "Wonder and Skepticism," 25.

10. Carson, *Sense of Wonder*, 42.

11. For a detailed and timely exploration of awe taking both philosophy and cognitive science into account, see De Cruz, *Wonderstruck.*

12. Burke, *A Philosophical Enquiry*, 303, 129.

13. Kant, *Critique of Judgment*, 76.

14. Kant, *Critique of Judgment*, 82.

15. Shapshay, "A Two-Tiered Theory of the Sublime."

16. Morton, "Poisoned Ground," 37.

17. Arcangeli et al., "Awe and the Experience of the Sublime," 21.

18. Joye and Verpooten, "An Exploration of the Functions of Religious Monumental Architecture," 53.

19. Joye and Verpooten, "An Exploration of the Functions of Religious Monumental Architecture," 57.

20. Valdesolo and Graham, "Awe, Uncertainty, and Agency Detection," 177.

21. Keltner and Haidt, "Approaching Awe," 312.

22. Murphy, "Sublime Dance of Mende Politics," 564.

23. Rampley, "Ethnographic Sublime."

24. Nye, *American Technological Sublime.*

25. Rampley, "Ethnographic Sublime," 261.

26. Rampley, "Ethnographic Sublime," 261.

27. Weibel, *A Sacred Vertigo*, 26.

28. Weibel, *A Sacred Vertigo*, 69.

29. Poole, Jeffares, and Penny, "Early Evolution," 880.

30. Angier, "Wonders of Blood."

31. Yong, *I Contain Multitudes.*

32. Muñoz-Delgado et al., "Moon Cycle Effects on Humans," 36.

33. Chakraborty, "Effects of Different Phases of the Lunar Month on Living Organisms," 255.

34. Williams, *Goldilocks Planet*.

Chapter 2. Ground to Sky

1. Thaden, *High, Wide and Frightened*, 138.
2. Wrangham, *Catching Fire*.
3. Kamra, "Role of Piggybacking."
4. Singer, *Like Sex with Gods*.
5. Gillispie, *Montgolfier Brothers*.
6. Davis, "Human Ascents into the Stratosphere," 198.
7. Tiberghien, Mukhamedjanova, and Xie, "Authenticity and Spectrality of Space Heritage," 1452.
8. Rodríguez-Laiz, "Wright Brothers"; Jamsari et al., "Ibn Firnas."
9. Adam, "De Dédale à Clément Ader"; Rodríguez-Laiz, "Wright Brothers."
10. Schlenoff, "Equivocal Success of the Wright Brothers."
11. Brandão, "Santos-Dumont Legacy," 6–8.
12. Corn, *Winged Gospel*, 4.
13. Corn, *Winged Gospel*, 13.
14. For more on the early military use of airplanes, see Morrow, *Great War in the Air*.
15. Van Riper, *Rockets and Missiles*.
16. Stern, "Robert Goddard and His Rockets."
17. Emme, *Robert H. Goddard*.
18. Neufeld, "Mittelbau Main Camp (aka Dora)."
19. Gimbel, "Project Paperclip."
20. Lichtblau, *Nazis Next Door*.
21. Bennett, "Blue Army and the Red Scare," 268.
22. Tiberghien, Mukhamedjanova, and Xie, "Authenticity and Spectrality of Space Heritage," 1452.
23. Pop, "Viewpoint: Space and Religion in Russia."
24. Shweder, "Illusions of 'Magical Thinking.'"
25. Weibel and Swanson, "Malinowski in Orbit," 53.
26. Weibel and Swanson, "Malinowski in Orbit," 56.

27. Weibel and Swanson, "Malinowski in Orbit," 56.

28. Hacker and Grimwood, *On the Shoulders of Titans*, 244.

29. Hacker and Grimwood, *On the Shoulders of Titan*, 249–50.

30. Collins, *Carrying the Fire*, 220.

31. National Archives and Records Administration, "America on the Move."

32. White, *Overview Effect*.

33. There has been some controversy about who said what, with some authors (and the official Apollo 8 transcript) indicating that Borman was the excited astronaut and Anders the one who wanted to stick to the plan. At a 2018 Apollo 8 anniversary event I attended at the Chicago Museum of Science and Industry, however, Anders, Borman, and Lovell were panelists, and both Anders and Lovell indicated that Borman discouraged taking the picture, which Anders did anyway. Space historian Chaikin, however, wrote an article explaining that Anders was, definitively, the awe-struck astronaut and Borman the wet blanket (Chaikin, "Who Took the Legendary Earthrise Photo?"). Finally, in an official NASA interview conducted in 2023, Anders confirmed Chaikin's claim that he (Anders) had been the one to spot the Earth and take its picture, so I'm going with that version of the story (YouTube, "Earthrise: A Conversation with Apollo 8 Astronaut Bill Anders").

34. YouTube, "NASA: Earthrise: The 45th Anniversary."

35. Chaikin, "Who Took the Legendary Earthrise Photo?"

36. White, *Overview Effect*, 3.

37. White, *Overview Effect*, 716.

38. Because my research participants came from a variety of backgrounds in a profession that is largely made up of heterosexual, white Christians, identifying participants outside of those parameters would risk exposing their identities in a way that violates my research permissions.

39. Weibel, "Black Ugliness and the Covering of Blue."

40. The biblical quotations to which he refers are Isaiah 40:22 and Job 26:8.

41. Massimino, *Spaceman*, 200.

Chapter 3. The Ultraview Effect

1. Haldane, "Possible Worlds," 286.

2. Lee et al., "Spaceflight Associated Neuro-Ocular Syndrome (SANS) and the Neuro-Ophthalmologic Effects of Microgravity."

3. Wells-Jensen, Schwartz, and Link, "3.A. Para-Astronauts."

4. Of the five recognized human senses of sight, hearing, smell, taste, and touch, sight seems to convey size and vastness the most directly, although hearing can, in some circumstances, convey scale and distance extremely well. Hearing, then, may also be a conduit for awe.

5. Feldman, "Constructing a Shared Bible Land," 355.

6. Weibel, *A Sacred Vertigo*, ix.

7. Kilgore, *Astrofuturism*; Gorman, *Dr Space Junk vs The Universe.*

8. Pantaleoni González et al., "Alma Catalogue of OB STARS."

9. Kraan-Korteweg and Lahav, "Universe behind the Milky Way."

10. Burke, *A Philosophical Enquiry*, 129.

11. Bell, Coan, and Oswald, "A Discussion on the Making of an EVA," 13.

Chapter 4. The Nature of Humility

1. NASA, "Who We Are."

2. Burke, *A Philosophical Enquiry*, 119.

3. Andres, *People Empowerment by Filipino Values*, 4.

4. Ioane, "Talanoa with Pasifika Youth and Their Families," 39.

5. Mashford-Pringle et al., "Weaving First Nations, Inuit, and Métis Principles and Values into Health Research Processes," 57.

6. Lambek, "Practice of Islamic Experts," 34.

7. Lee, *Dobe Ju/'hoansi*, 57.

8. Trivers, "Reciprocal Altruism."

9. Wilson and Wilson, "Rethinking the Theoretical Foundation of Sociobiology."

10. Henrich, *Weirdest People in the World.*

11. Chen et al., "Chinese and American Individuals' Mate Selection Criteria."

12. Lee and Ashton, *H Factor of Personality.*

13. Krumrei-Mancuso et al., "Links between Intellectual Humility and Acquiring Knowledge," 155.

14. Tetlock, *Expert Political Judgment.*

15. Encyclopædia Britannica, s.v. "Confirmation Bias."

16. Porter et al., "Predictors and Consequences of Intellectual Humility," 531.

17. Kolb, *Experiential Learning.*

18. Bloom et al., *Handbook I: Cognitive Domain.*

19. Bloom et al., *Handbook I: Cognitive Domain*; Encyclopædia Britannica, s.v. "Bloom's Taxonomy."

20. See Vatican Observatory, "Frequently Asked Questions." https://www.vaticanobservatory.va/en/contact/faq.

21. NASA, "WMAP Big Bang Theory."

Chapter 5. Navigating Humility

1. Einstein, letter to Queen Elisabeth of Belgium, September 1932, quoted in Dukas and Hoffmann, *Einstein: The Human Side*, 38.

2. Van Tongeren, *Humble.*

3. Worth and Reeves, "'I Am Almost the Middle-Class White Man, Aren't I?'"

4. Van Tongeren, *Humble.*

5. St. Laurent, "Light and Silence."

6. Fox, *George Fox's Journal.*

7. McAleer, "Whittier's Quest for Humility"; Birch, "Quaker Plain Speech."

8. St. Augustine and Sparrow-Simpson, *Letters of St. Augustine*, 53.

9. Roth, "Towards a Definition of Humility," 8–9.

10. Cook, *A History of the Muslim World.*

11. Reynolds, *Qur'ān in Its Historical Context.*

12. Smith, "Eaters, Food, and Social Hierarchy in Ancient India."

13. *Devi Bhagavata Purana.*

14. Smith, *Essence of Buddhism.*

15. Smith, *Essence of Buddhism.*

16. Suzuki et al., *Zen Mind, Beginner's Mind*, xiv–2.

17. Murphy-Shigematsu, "Vulnerability and Beginner's Mind," 399.

18. Chirico and Yaden, *Awe*, 223.

19. McCloskey, Washburn, and Felch, "Intuitive Physics."

20. Crapanzano, "'Strange Trajectories.'"

21. Norris and Norris, "Why Are There Seven Sisters?"

22. Faulkner, *Falling Flat.*

23. Jones, Adams, and Mayoh, "Motivated Ignorance and Social Identity Threat."

24. Rowbotham, *Zetetic Astronomy*.

25. Fernbach and Bogard, "Conspiracy Theory as Individual and Group Behavior."

26. Mitchell and Williams, *Way of the Explorer*, 70, 4, 69.

27. *People*, "Edgar Mitchell's Strange Voyage."

Chapter 6. From Awe to Action

1. Quoted in Sarma, *Mental Resilience*, 61.

2. Bronston, "Roll over, Aristotle, Nature Doesn't Always Hate a Vacuum"; Barnett et al., "Spontaneous Drying."

3. McCrae, "Openness to Experience," 41.

4. Block, "A Contrarian View."

5. Gurven et al., "How Universal Is the Big Five?"

6. Merriam-Webster, "Exploring Definition & Meaning."

7. Glausiusz, "What Drove Homo Erectus out of Africa?"

8. Białek-Dratwa et al., "Neophobia."

9. Plato, Benjamin Jowett, and J. D. Kaplan, *Dialogues of Plato*.

10. Mackenzie, "Virtues of Socratic Ignorance."

11. Boyle, "Pure of Heart."

12. Koyre, "Galileo and Plato"; Bardi, "Copernicus and Axiomatics"; and Goddu, *Copernicus and the Aristotelian Tradition*.

13. Omar, "Ibn al-Haytham's Theory of Knowledge."

14. Gilbert, *De Magnete*.

15. Miller and Jalobeanu, *Cambridge History of Philosophy of the Scientific Revolution*.

16. Bhattacharya, "Post-Enlightenment Exploration and the Aesthetic of Information."

17. Tätte et al., "Ami and Yami Aborigines of Taiwan."

18. Benham et al., *'Ike Ulana Lau Hala*.

19. Kirch, "Peopling of the Pacific."

20. Chambers, *Joseph Banks and the British Museum*.

21. Jenkinson, "Heroic and Problematic Legacy of Joseph Banks."

22. Chambers, *Joseph Banks and the British Museum*.

23. Obeyesekere, *Apotheosis of Captain Cook*; Ashley, "How Navigators Think."

24. O'Sullivan, "Annexation."

25. Theilmann, "Medieval Pilgrims."

26. Bercovitch and Sollors, "Pilgrims and Puritans."

27. Weinberg, *Manifest Destiny: A Study of Nationalist Expansionism.*

28. Stephanson and Foner, *Manifest Destiny: American Expansionism.*

29. Newell, *Destined for the Stars.*

30. Newell, *Destined for the Stars.*

31. Traphagan, "Religion, Science, and Space Exploration."

32. Tavares et al., "Ethical Exploration."

33. Nesvold, *Off-Earth*, 19.

34. Nesvold, *Off-Earth*, 18.

35. Feehly, "Giant Space Habitats."

36. Johnson-Schwartz, Billings, and Nesvold, "An Introduction to Reclaiming Space"; Nesvold, *Off-Earth*, 19.

37. Tavares et al., "Ethical Exploration."

38. Tavares et al., "Ethical Exploration"; Johnson-Schwartz and Milligan, "Some Ethical Constraints."

39. Tavares et al., "Ethical Exploration."

40. Tavares et al., "Ethical Exploration"; Weibel, "Sherpas on the Moon."

41. Johnson-Schwartz and Milligan, "Some Ethical Constraints"; Wood, Muniyappa, and Reed, "Opportunities."

42. Newell, *Destined for the Stars;* Swanson, "New Frontier," 592; and Beattie, *Taking Science to the Moon.*

43. See Johnson-Schwartz and Milligan, "Some Ethical Constraints"; Tavares et al., "Ethical Exploration"; Gorman, "Space Debris"; and Weibel, "Sacred Moon."

Chapter 7. Stepping Together into Uncertainty

1. Quoted in Mosher, "SpaceX Just Docked the First Commercial Spaceship."

2. Russo and Bailey, "Deeper Insights into the Solar Eclipse Experience."

3. Kline, "International Space Station Retires Soon."

4. Hōkūle'a, "History."

5. Lepere, "Pius 'Mau' Piailug."

6. Thompson, "Reflections on Mau Piailug."

7. McAvoy, "How a Canoe Helped Turn Hawaiian Culture into a Source of Pride."

8. McAvoy, "How a Canoe Helped Turn Hawaiian Culture into a Source of Pride."

9. Eddie Aikau is also honored by the periodic Eddie Aikau Big Wave Invitational (aka "the Eddie"), a surfing contest held at Oahu's Waimea Bay during conditions when the waves are more than 30 feet high.

10. Carson, *Sea around Us*, 14.

11. Agarwal, "Deep Sea."

12. Polmar and Mathers, *Opening the Great Depths.*

13. Piccard and Dietz, *Seven Miles Down.*

14. Harris, "Titan Submersible Disaster."

15. Yang, Liu, and Liao, "Manned Submersibles."

16. Weibel, "Highs and Lows of Extreme Tourism."

17. FSC Indigenous Foundation, "Hindou Oumarou Ibrahim."

18. Crawhall, *Pastoralists Seek Water and Peace in Chad.*

19. McCool, "'Grandmothers Are Our Weather App.'"

20. *Wired*, "How Mapping Indigenous Knowledge Is Helping Nomadic Communities."

21. FSC Indigenous Foundation, "Hindou Oumarou Ibrahim."

22. *Wired*, "How Mapping Indigenous Knowledge Is Helping Nomadic Communities."

Opening Vistas

1. Greene, "Universe on a String."

Bibliography

Adam, Jean-Pierre. "De Dédale à Clément Ader." *Alliage: Culture-Science-Technique* 6 (1990): 35–42.

Agarwal, Neal. "The Deep Sea." Neil.fun. Accessed July 31, 2024. www.neil.fun/deep-sea.

Andres, Tomas D. *People Empowerment by Filipino Values.* Rex Book Store, 1998.

Angier, Natalie. "The Wonders of Blood." *New York Times,* October 20, 2008.

Arcangeli, Margherita, Marco Sperduti, Amélie Jacquot, Pascale Piolino, and Jérôme Dokic. "Awe and the Experience of the Sublime: A Complex Relationship." *Frontiers in Psychology* 11 (June 16, 2020). https://doi.org/10.3389/fpsyg.2020.01340.

Ashley, S. "How Navigators Think: The Death of Captain Cook Revisited." *Past & Present* 194, no. 1 (February 1, 2007): 107–37. https://doi.org/10.1093/pastj/gtl019.

Augustine, St., and W. J. Sparrow-Simpson. *The Letters of St. Augustine.* Macmillan, 1919.

Aveni, Anthony. *In the Shadow of the Moon: The Science, Magic, and Mystery of Solar Eclipses.* Yale University Press, 2017.

Bardi, Alberto. "Copernicus and Axiomatics." In *Handbook of the History and Philosophy of Mathematical Practice*, December 20, 2022. https://doi.org/10.1007/978-3-030-19071-2_110-1.

Barnett, J. Wesley, Matthew R. Sullivan, Joshua A. Long, Du Tang, Thong Nguyen, Dor Ben-Amotz, Bruce C. Gibb, and Henry S. Ashbaugh.

"Spontaneous Drying of Non-Polar Deep-Cavity CAVITAND Pockets in Aqueous Solution." *Nature Chemistry* 12, no. 7 (May 18, 2020): 589–94. https://doi.org/10.1038/s41557-020-0458-8.

BBC Radio 4. "Sideways, a New Frontier: 4. With the Gods." January 2, 1970. https://www.bbc.co.uk/programmes/m0021j90.

Beattie, Donald A. *Taking Science to the Moon: Lunar Experiments and the Apollo Program*. The Johns Hopkins University Press, 2003.

Bell, Ernest, David Coan, and David Oswald. "A Discussion on the Making of an EVA: What It Really Takes to Walk in Space." SpaceOps 2006 Conference, June 19, 2006. https://doi.org/10.2514/6.2006-5570.

Benham, Maenette K. P. Ah Nee, C. Kurt Dewhurst, Timothy Gallaher, Betty Lou Kam, Lia O'Neill M. A. Keawe, Elizabeth Maluihi Lee, and Kaiwipunikauikawēkiu Lipe. *'Ike Ulana Lau Hala: The Vitality and Vibrancy of Lau Hala Weaving Traditions in Hawaiʻi*. University of Hawaiʻi Press, 2014.

Bennett, Jeffrey S. "The Blue Army and the Red Scare: Politics, Religion, and Cold War Paranoia." *Politics, Religion & Ideology* 16, no. 2–3 (April 3, 2015): 263–81. https://doi.org/10.1080/21567689.2015.1084927.

Bercovitch, Sacvan, and Werner Sollors. "Pilgrims and Puritans and the Myth of the Promised Land." In *The Myths That Made America: An Introduction to American Studies*. CPI—Clausen & Bosse, 2014.

Bhattacharya, Diganta. "Post-Enlightenment Exploration and the Aesthetic of Information: Curious with a Purpose." *Rupkatha Journal on Interdisciplinary Studies in Humanities* 13, no. 1 (March 28, 2021). https://doi.org/10.21659/rupkatha.v13n1.16.

Białek-Dratwa, Agnieszka, Elżbieta Szczepańska, Dorota Szymańska, Mateusz Grajek, Karolina Krupa-Kotara, and Oskar Kowalski. "Neophobia: A Natural Developmental Stage or Feeding Difficulties for Children?" *Nutrients* 14, no. 7 (April 6, 2022): 1521. https://doi.org/10.3390/nu14071521.

Birch, Barbara M. "Quaker Plain Speech: A Policy of Linguistic Divergence." *International Journal of the Sociology of Language* 116, no. 1 (January 1, 1995). https://doi.org/10.1515/ijsl.1995.116.39.

Block, Jack. "A Contrarian View of the Five-Factor Approach to Personality Description." *Psychological Bulletin* 117, no. 2 (1995): 187–215. https://doi.org/10.1037//0033-2909.117.2.187.

Bloom, Benjamin S., David R. Krathwohl, Walker H. Hill, E. J. Furst, and Max D. Englehart. *Handbook I: Cognitive Domain.* David McKay, 1956.

Boyle, Marjorie O'Rourke. "Pure of Heart: From Ancient Rites to Renaissance Plato." *Journal of the History of Ideas* 63, no. 1 (January 2002): 41–62. https://doi.org/10.1353/jhi.2002.0002.

Brandão, Mauricio Pazini. "The Santos-Dumont Legacy to Aeronautics." In *31st Conference of the National Council of the Aeronautical Sciences.* Belo Horizonte, Brazil, 2018.

Bronston, Barri. "Roll over, Aristotle, Nature Doesn't Always Hate a Vacuum." *Tulane University News*, June 15, 2020. https://news.tulane.edu/pr/roll-over-aristotle-nature-doesn%E2%80%99t-always-hate-vacuum.

Burke, Edmund. *A Philosophical Enquiry into the Origin of Our Ideas of the Sublime and Beautiful.* J. Dodsley, 1767.

Burnet, Thomas. *The Sacred Theory of the Earth.* True Sign Publishing, 2023.

Carrive, Paulette. "Le sublime dans l'esthétique de Kant." *Revue d'histoire littéraire de la France* 86, no. 1 (1986): 71–85.

Carson, Rachel. *The Sea around Us.* Golden Press, 1958.

———. *Sense of Wonder: A Celebration of Nature for Parents and Children.* Harpercollins, 1965.

Chaikin, Andrew. "Who Took the Legendary Earthrise Photo from Apollo 8?" Smithsonian.com, January 1, 2018. https://www.smithsonianmag.com/science-nature/who-took-legendary-earthrise-photo-apollo-8-180967505.

Chakraborty, Ujjwal. "Effects of Different Phases of the Lunar Month on Living Organisms." *Biological Rhythm Research* 51, no. 2 (October 8, 2018): 254–82. https://doi.org/10.1080/09291016.2018.1526502.

Chambers, Neil. *Joseph Banks and the British Museum: The World of Collecting, 1770–1830.* Routledge, 2018.

Chen, Ruoxi, Jason P. Austin, John K. Miller, and Fred P. Piercy. "Chinese and American Individuals' Mate Selection Criteria: Updates, Modifications, and Extensions." *Journal of Cross-Cultural Psychology* 46, no. 1 (2015): 101–18.

Chirico, Alice, and Yaden, D. B. *Awe: A Self-Transcendent and Sometimes Transformative Emotion.* Springer, 2018.

Clark, Ronald. *Einstein: The Life and Times; An Illustrated Biography.* H. N. Abrams, 1971.

Collins, Michael. *Carrying the Fire: An Astronaut's Journeys; 50th Anniversary Edition*. Farrar, Straus and Giroux, 2019.

Cook, Michael. *A History of the Muslim World: From Its Origins to the Dawn of Modernity*. Princeton University Press, 2024.

Corn, Joseph. *The Winged Gospel*. Oxford University Press, 1983.

Crapanzano, Francesco. "'Strange Trajectories.'" *Transversal: International Journal for the Historiography of Science* 5 (December 9, 2018). https://doi.org/10.24117/2526-2270.2018.i5.06.

Crawhall, Nigel (IPACC). "Pastoralists Seek Water and Peace in Chad: Account of a Participatory Mapping Exercise in the Sahel." PPgis.net Blog (PPGIS/PGIS Blog), August 16, 2012. Posted by Giacomo Rambaldi. https://participatorygis.blogspot.com/2012/08/pastoralists-seek-water-and-peace-in.html.

Cunningham, George. "New Disneyland Ride Previewed." *Orange County Register*, April 29, 1977.

Daniel, Mitchel J., Laura Koffinas, and Kimberly A. Hughes. "Mating Preference for Novel Phenotypes Can Be Explained by General Neophilia in Female Guppies." *The American Naturalist* 196, no. 4 (October 1, 2020): 414–28. https://doi.org/10.1086/710177.

Davis, Watson. "Human Ascents into the Stratosphere." *Bulletin of the American Meteorological Society* 15, nos. 8–9 (August 1, 1934): 198–202. https://doi.org/10.1175/1520-0477-15.8-9.198.

Day, Rachel L., Rebecca L. Coe, Jeremy R. Kendal, and Kevin N. Laland. "Neophilia, Innovation and Social Learning: A Study of Intergeneric Differences in Callitrichid Monkeys." *Animal Behaviour* 65, no. 3 (March 2003): 559–71. https://doi.org/10.1006/anbe.2003.2074.

De Cruz, Helen. *Wonderstruck: How Wonder and Awe Shape the Way We Think*. Princeton University Press, 2024.

Devi Bhagavata Purana. Translated by Swami Vijnanananda. Parimal, 1998.

Einstein, Albert. *Albert Einstein, The Human Side: Glimpses from His Archives*. Edited by Helen Dukas and Banesh Hoffmann. Princeton University Press, 1981.

Elvis, Martin, Alanna Krolikowski, and Tony Milligan. "Space Resources: Physical Constraints, Policy Choices, and Ethical Considerations." *In-Space Manufacturing and Resources*, July 2022, 369–88. https://doi.org/10.1002/9783527830909.ch20.

Emme, Eugene M. *Robert H. Goddard: World Rocket Pioneer*. Office of Technical Information and Educational Programs, National Aeronautics and Space Administration, 1960.

Faulkner, Danny R. *Falling Flat: A Refutation of Flat Earth Claims*. Master Books, 2019.

Feehly, Conor. "Giant Space Habitats: Are They Science Fiction or the Future?" *Discover Magazine*, July 14, 2023. https://www.discovermagazine.com/the-sciences/giant-space-habitats-are-they-science-fiction-or-the-future.

Feldman, Jackie. "Constructing a Shared Bible Land: Jewish Israeli Guiding Performances for Protestant Pilgrims." *American Ethnologist* 34, no. 2 (May 2007): 351–74. https://doi.org/10.1525/ae.2007.34.2.351.

Fernbach, Philip M., and Jonathan E. Bogard. "Conspiracy Theory as Individual and Group Behavior: Observations from the Flat Earth International Conference." *Topics in Cognitive Science* 16, no. 2 (May 18, 2023): 187–205. https://doi.org/10.1111/tops.12662.

Fox, George. *George Fox's Journal*. Kessinger Publishing, 2009.

FSC Indigenous Foundation. "Hindou Oumarou Ibrahim." January 14, 2025. https://www.fscindigenousfoundation.org/teams/hindou-oumarou-ibrahim.

Fung, Katherine. "Why Can't Politicians Ever Admit They've Been Wrong?" *Newsweek*, October 25, 2024. https://www.newsweek.com/why-cant-politicians-ever-admit-theyve-been-wrong-1973022.

Garner, Rob. "Dr. Robert H. Goddard, American Rocketry Pioneer." NASA, August 3, 2017. https://www.nasa.gov/centers/goddard/about/history/dr_goddard.html.

Gilbert, William. *De Magnete*. Dover, 2013.

Gillispie, Charles Coulston. *The Montgolfier Brothers and the Invention of Aviation 1783–1784: With a Word on the Importance of Ballooning for the Science of Heat and the Art of Building Railroads*. Princeton University Press, 2014.

Gimbel, John. "Project Paperclip: German Scientists, American Policy, and the Cold War." *Diplomatic History* 14, no. 3 (July 1990): 343–66. https://doi.org/10.1111/j.1467-7709.1990.tb00095.x.

Glausiusz, Josie. "What Drove Homo Erectus out of Africa?" *Smithsonian Magazine*, October 19, 2021. https://www.smithsonianmag.com/science-nature/what-drove-homo-erectus-out-of-africa-180978881.

Goddu, André. *Copernicus and the Aristotelian Tradition: Education, Reading, and Philosophy in Copernicus's Path to Heliocentrism*. Brill, 2010.

Gorman, Alice. *Dr Space Junk vs The Universe: Archaeology and the Future*. MIT Press, 2019.

———. "Space Debris, Space Situational Awareness and Cultural Heritage Management in Earth Orbit." *Commercial and Military Uses of Outer Space*, 2021, 133–51. https://doi.org/10.1007/978-981-15-8924-9_10.

Grant, Adam. *Think Again: The Power of Knowing What You Don't Know*. Penguin Books, 2023.

Greene, Brian. "The Universe on a String." *New York Times*, October 20, 2006.

Gurven, Michael, Christopher von Rueden, Maxim Massenkoff, Hillard Kaplan, and Marino Lero Vie. "How Universal Is the Big Five? Testing the Five-Factor Model of Personality Variation among Forager–Farmers in the Bolivian Amazon." *Journal of Personality and Social Psychology* 104, no. 2 (February 2013): 354–70. https://doi.org/10.1037/a0030841.

Hacker, Barton C., and James M. Grimwood. *On the Shoulders of Titans: A History of Project Gemini*. NASA, Scientific and Technical Information, 1977.

Haldane, J. B. S. "Possible Worlds." In *Possible Worlds: And Other Essays*. Chatto and Windus, 1932.

Harris, Mark. "The Titan Submersible Disaster Shocked the World. The Inside Story Is More Disturbing than Anyone Imagined." *Wired*, June 11, 2024. https://www.wired.com/story/titan-submersible-disaster-inside-story-oceangate-files.

Hasbrook, Pete, Julie A. Robinson, Judy Brown Tate, Tracy Thumm, Luchino Cohen, Isabelle Marcil, and Lina De Parolis. "Benefits of International Collaboration on the International Space Station." In *International Astronautical Congress (IAC 2017) JSC-CN-40449-1. 2017, JSC-CN-40449-1*. Johnson Space Center, 2017.

Henrich, Joseph Patrick. *The Weirdest People in the World: How the West Became Psychologically Peculiar and Particularly Prosperous*. Farrar, Straus and Giroux, 2021.

Higginbotham, Adam. *Challenger: A True Story of Heroism and Disaster on the Edge of Space*. Avid Reader Press, 2024.

Hōkūle'a. "History." Accessed May 21, 2025. https://hokulea.com/history.

Howard, Mindy. *Blast Off! Train Like an Astronaut for Success on Earth*. Pumbo, 2019.

Institute for Experiential Learning. "What Is Experiential Learning?" December 27, 2023. https://experientiallearninginstitute.org/what-is-experiential-learning.

Institute of the Environment and Sustainability at UCLA. "Hindou Oumarou Ibrahim." May 20, 2025. https://www.ioes.ucla.edu/person/hindou-oumarou.

Inzani, Emma L., Laura A. Kelley, and Neeltje J. Boogert. "Object Neophilia in Wild Herring Gulls in Urban and Rural Locations." *Journal of Avian Biology* 2023, no. 1–2 (December 15, 2022). https://doi.org/10.1111/jav.03028.

Ioane, Julia. "Talanoa with Pasifika Youth and Their Families." *New Zealand Journal of Psychology* 46, no. 3 (2017): 38–45.

Ioannidis, Alexander G., Javier Blanco-Portillo, Karla Sandoval, Erika Hagelberg, Carmina Barberena-Jonas, Adrian V. Hill, and Juan Esteban Rodríguez-Rodríguez. "Paths and Timings of the Peopling of Polynesia Inferred from Genomic Networks." *Nature* 597, no. 7877 (September 22, 2021): 522–26. https://doi.org/10.1038/s41586-021-03902-8.

IONS. "Institute of Noetic Sciences (IONS)." June 14, 2024. https://noetic.org.

Jahan, S. M. Mukarram. "Religious Experience of Awe/Fear in the Qur'ān and Its Parallels in the Bible." Nd Mushirul Haque National Seminar of Research Scholars in Islamic Studies. Lecture, 2023.

Jamsari, Ezad Azraai, Mohd Aliff Mohd Nawi, Adibah Sulaiman, Roziah Sidik, Zanizam Zaidi, and M. Z. A. H. Ashari. "Ibn Firnas and His Contribution to the Aviation Technology of the World." *Advances in Natural and Applied Sciences* 7, no. 1 (2013): 73–78.

Jenkinson, Clay. "The Heroic and Problematic Legacy of Joseph Banks." *Listening To America*, June 11, 2024. https://ltamerica.org/the-heroic-and-problematic-legacy-of-joseph-banks.

Johnson-Schwartz, J. S., and Tony Milligan. "Some Ethical Constraints on Near-Earth Resource Exploitation." *Yearbook on Space Policy, 2017*, 227–39. https://doi.org/10.1007/978-3-7091-4860-0_10.

Johnson-Schwartz, J. S., Linda Billings, and Erika Nesvold. "An Introduction to Reclaiming Space." In *Reclaiming Space: Progressive and Multicultural Visions of Space Exploration*. Oxford University Press, 2023.

Jones, Ian, Andrew Adams, and Joanne Mayoh. "Motivated Ignorance and Social Identity Threat: The Case of the Flat Earth." *Social Identities* 29, no. 1 (January 2, 2023): 79–94. https://doi.org/10.1080/13504630.2023.2208033.

Joye, Yannick, and Jan Verpooten. "An Exploration of the Functions of Religious Monumental Architecture from a Darwinian Perspective." *Review of General Psychology* 17, no. 1 (March 2013): 53–68. https://doi.org/10.1037/a0029920.

Kahneman, Daniel. *Thinking, Fast and Slow*. Penguin Books, 2024.

Kamra, Aakriti. "The Role of Piggy Backing in Technological Leapfrogging in Developing Countries." *International Development Planning Review* 23, no. 1 (2024): 55–64.

Kant, Immanuel. *Critique of Judgment*. Macmillan, 1914. Originally published in 1790.

Kappeler, Peter M. *Cooperation in Primates and Humans: Mechanisms and Evolution*. Springer, 2006.

Keltner, Dacher. *Awe: The New Science of Everyday Wonder and How It Can Transform Your Life*. Penguin, 2023.

Keltner, Dacher, and Jonathan Haidt. "Approaching Awe, a Moral, Spiritual, and Aesthetic Emotion." *Cognition and Emotion* 17, no. 2 (January 2003): 297–314. https://doi.org/10.1080/02699930302297.

Kilgore, De Witt Douglas. *Astrofuturism: Science, Race, and Visions of Utopia in Space*. University of Pennsylvania Press, 2010.

Kirch, Patrick Vinton. "Peopling of the Pacific: A Holistic Anthropological Perspective." *Annual Review of Anthropology* 39 (2010): 131–48.

Kline, Kaity. "The International Space Station Retires Soon. NASA Won't Run Its Future Replacement." NPR, February 21, 2024. https://www.npr.org/2024/02/21/1232639289/international-space-station-retirement-space-stations-future.

Kolb, David A. *Experiential Learning: Experience as the Source of Learning and Development*. Pearson Education, 2022.

Koyre, Alexandre. "Galileo and Plato." *Journal of the History of Ideas* 4, no. 4 (October 1943): 400. https://doi.org/10.2307/2707166.

Kraan-Korteweg, Rene C., and Ofer Lahav. "The Universe behind the Milky Way." *Astronomy and Astrophysics Review* 10, no. 3 (September 7, 2000): 211–61. https://doi.org/10.1007/s001590000011.

Krumrei-Mancuso, Elizabeth J., Megan C. Haggard, Jordan P. LaBouff, and Wade C. Rowatt. "Links between Intellectual Humility and Acquiring Knowledge." *The Journal of Positive Psychology* 15, no. 2 (February 14, 2019): 155–70. https://doi.org/10.1080/17439760.2019.1579359.

Lambek, Michael. "The Practice of Islamic Experts in a Village on Mayotte." *Journal of Religion in Africa* 20, no. 1 (1990): 20–40. https://doi.org/10.1163/157006690x00033.

Lambert, Jonathan. "It's America's 'Manifest Destiny' to Plant a Flag on Mars, Trump Says." NPR, January 20, 2025. https://www.npr.org/2025/01/20/g-s1-43861/mars-nasa-manifest-destiny-trump-inauguration.

Lee, Andrew G., Thomas H. Mader, C. Robert Gibson, William Tarver, Pejman Rabiei, Roy F. Riascos, Laura A. Galdamez, and Tyson Brunstetter. "Spaceflight Associated Neuro-Ocular Syndrome (SANS) and the Neuro-Ophthalmologic Effects of Microgravity: A Review and an Update." *npj Microgravity* 6, no. 1 (February 7, 2020). https://doi.org/10.1038/s41526-020-0097-9.

Lee, Kibeom, and Michael C. Ashton. *The H Factor of Personality: Why Some People Are Manipulative, Self-Entitled, Materialistic, and Exploitive—And Why It Matters for Everyone*. Wilfrid Laurier University Press, 2012.

Lee, Richard B. *The Dobe Ju/'hoansi*. Harcourt Brace College Publishers, 1993.

Lepere, Imogen. "Pius 'Mau' Piailug: Master Navigator of Micronesia." *JSTOR Daily*, August 21, 2023. https://daily.jstor.org/pius-mau-piailug-master-navigator-of-micronesia.

Lichtblau, Eric. *The Nazis Next Door: How America Became a Safe Haven for Hitler's Men*. Mariner Books, 2015.

Longinus. *Peri hypsous: Or Dionysius Longinus of the Height of Eloquence; Rendred out of the Originall by J.H. ESQ.* Translated by John Hall. Printed by Roger Daniel for Francis Eaglesfield at the Marygold in Pauls Church-yard, 1652.

Lorente, Araceli Giménez. "The Origin of Art: An Approximation through Archaeological Evidences." *Sociology and Anthropology* 8, no. 3 (May 2020): 82–91. https://doi.org/10.13189/sa.2020.080304.

Mackenzie, Mary Margaret. "The Virtues of Socratic Ignorance." *The Classical Quarterly* 38, no. 2 (December 1988): 331–50. https://doi.org/10.1017/s0009838800037009.

Mallove, Eugene F. "Science, Scientism, and Meaning." *Infinite Energy* 30 (2000): 4–5.

Mashford-Pringle, Angela, Claire Hiscock, Emma Janet Rice, and Bryanna Scott. "Weaving First Nations, Inuit, and Métis Principles and Values into Health Research Processes." *Journal of Clinical Epidemiology* 160 (August 2023): 54–60. https://doi.org/10.1016/j.jclinepi.2023.05.012.

Massimino, Mike. *Spaceman: An Astronaut's Unlikely Journey to Unlock the Secrets of the Universe*. Simon & Schuster, 2016.

McAleer, John J. "Whittier's Quest for Humility." *Bulletin of Friends' Historical Association* 50, no. 1 (March 1961): 31–45. https://doi.org/10.1353/qkh.1961.a395377.

McAvoy, Audrey. "How a Canoe Helped Turn Hawaiian Culture into a Source of Pride and Even Influenced Hollywood." *Honolulu Civil Beat*, March 8, 2025. https://www.civilbeat.org/2025/03/how-a-canoe-helped-turn-Hawaiian-culture-into-a-source-of-pride-and-even-influenced-hollywood.

McCloskey, Michael, Allyson Washburn, and Linda Felch. "Intuitive Physics: The Straight-down Belief and Its Origin." *Journal of Experimental Psychology: Learning, Memory, and Cognition* 9, no. 4 (1983): 636–49. https://doi.org/10.1037//0278-7393.9.4.636.

McCool, Alice. "'Grandmothers Are Our Weather App': New Maps and Local Knowledge Power Chad's Climate Fightback." *The Guardian*, August 25, 2022. https://www.theguardian.com/global-development/2022/aug/25/new-maps-and-local-knowledge-power-chad-climate-fightback-hindou-oumarou-ibrahim.

McCrae, Robert R. "Openness to Experience as a Basic Dimension of Personality." *Imagination, Cognition and Personality* 13, no. 1 (September 1993): 39–55. https://doi.org/10.2190/h8h6-qykr-keu8-gaq0.

Meyerowitz, Joanne J. *Not June Cleaver: Women and Gender in Postwar America, 1945–1960*. Temple University Press, 2006.

Miller, David Marshall, and Dana Jalobeanu. *The Cambridge History of Philosophy of the Scientific Revolution*. Cambridge University Press, 2022.

Mitchell, Edgar D., and Dwight Arnan Williams. *The Way of the Explorer: An Apollo Astronaut's Journey through the Material and Mystical Worlds*. New Page Books, 2008.

Morrow, John H. *The Great War in the Air: Military Aviation from 1909 to 1921*. University of Alabama Press, 2009.
Morton, Timothy. "Poisoned Ground: Art and Philosophy in the Time of Hyperobjects." *Symplokē* 21, no. 1–2 (2013): 37–50. https://doi.org/10.5250/symploke.21.1-2.0037.
Mosher, Dave. "SpaceX Just Docked the First Commercial Spaceship Built for Astronauts to the International Space Station—What NASA Calls a 'Historic Achievement.'" *Business Insider*, March 3, 2019. https://www.businessinsider.com/nasa-astronauts-connect-spacex-crew-dragon-spaceship-international-space-station-2020-5.
Muñoz-Delgado, Jairo, Ana María Santillán-Doherty, Ricardo Mondragón-Ceballos, and Hans G. Erkert. "Moon Cycle Effects on Humans: Myth or Reality?" *Salud Mental* 23, no. 6 (2000): 33–39.
Murphy, William P. "The Sublime Dance of Mende Politics: An African Aesthetic of Charismatic Power." *American Ethnologist* 25, no. 4 (November 1998): 563–82. https://doi.org/10.1525/ae.1998.25.4.563.
Murphy-Shigematsu, Stephen. "Vulnerability and Beginner's Mind." *The Clinical Teacher* 11, no. 5 (August 2014): 399–400.
Nakayama, Masataka, Yuki Nozaki, Pamela M. Taylor, Dacher Keltner, and Yukiko Uchida. "Individual and Cultural Differences in Predispositions to Feel Positive and Negative Aspects of Awe." *Journal of Cross-Cultural Psychology* 51, no. 10 (September 27, 2020): 771–93. https://doi.org/10.1177/0022022120959821.
NASA. "Transcripts of Lro_apollo_earthrise_ipod_lg." Accessed May 15, 2024. https://svs.gsfc.nasa.gov/vis/a010000/a010900/a010961/lro_apollo_earthrise_transcript.html.
———. "Who We Are." Accessed June 5, 2024. www.jpl.nasa.gov/who-we-are.
———. "WMAP Big Bang Theory." Accessed June 12, 2024. https://wmap.gsfc.nasa.gov/universe/bb_theory.html.
National Archives and Records Administration. "America on the Move: Crew of Apollo 8—A View from Lunar Orbit, 1968." Accessed May 10, 2024. https://www.archives.gov/exhibits/eyewitness/html.php?section=25.
Nausicaá Boulogne-sur-Mer. "Fosse des Mariannes, 11 000 mètres sous les mers." March 29, 2024. www.nausicaa.fr/fr/le-mag-ocean/fosse-des-mariannes-11-000-metres-sous-les-mers.

Nesvold, Erika. *Off-Earth: Ethical Questions and Quandaries for Living in Outer Space*. MIT Press, 2023.

Neufeld, Michael J. "Mittelbau Main Camp (aka Dora)." In *The United States Holocaust Memorial Museum Encyclopedia of Camps and Ghettos, 1933–1945*. Vol. 1, *Early Camps, Youth Camps, and Concentration Camps and Subcamps under the SS-Business Administration Main Office (WVHA)*. Edited by Geoffrey P. Megargee. Indiana University Press, 2009.

Newell, Catherine L. *Destined for the Stars: Faith, the Future, and America's Final Frontier.* University of Pittsburgh Press, 2019.

Norris, Ray P., and Barnaby R. Norris. "Why Are There Seven Sisters?" *Advancing Cultural Astronomy*, 2021, 223–35. https://doi.org/10.1007/978-3-030-64606-6_11.

Nye, David E. *American Technological Sublime*. MIT Press, 2007.

Obeyesekere, Gananath. *The Apotheosis of Captain Cook: European Mythmaking in the Pacific*. Princeton University Press, 1992.

Omar, Saleh. "Ibn al-Haytham's Theory of Knowledge and Its Significance for Later Science." *Arab Studies Quarterly* (1979): 67–82.

O'Sullivan, John L. "Annexation." *United States Magazine and Democratic Review* 17, no. 85 (1845): 5–10.

Packer, Alex. *Wise Highs*. Turtleback Books, 2006.

Pantaleoni González, M., J. Maíz Apellániz, R. H. Barbá, and B. Cameron Reed. "The Alma Catalogue of OB STARS—II. A Cross-Match with Gaia DR2 and an Updated Map of the Solar Neighbourhood." *Monthly Notices of the Royal Astronomical Society* 504, no. 2 (March 19, 2021): 2968–82. https://doi.org/10.1093/mnras/stab688.

Paul, Heike. *Myths That Made America: An Introduction to American Studies*. Transcript, 2014.

Paul, Richard, and Steven Moss. *We Could Not fail: The First African Americans in the Space Program*. University of Texas Press, 2021.

People. "Edgar Mitchell's Strange Voyage." April 8, 1974.

Piccard, Jacques, and Robert S. Dietz. *Seven Miles Down: The Story of the Bathyscaph Trieste*. Scientific Book Club, 1963.

Plato, Benjamin Jowett, and J. D. Kaplan. *Dialogues of Plato: Jowett Translation*. Pocket Books, 2001.

Polmar, Norman, and Lee J. Mathers. *Opening the Great Depths: The Bathyscaph Trieste and Pioneers of Undersea Exploration*. Naval Institute Press, 2021.

Poole, Anthony, Daniel Jeffares, and David Penny. "Early Evolution: Prokaryotes, the New Kids on the Block." *BioEssays* 21, no. 10 (September 23, 1999): 880-89. https://doi.org/10.1002/(sici)1521-1878(199910)21:10{{lt;}}880::aid-bies11{{gt;}}3.0.co;2-p.

Pop, Virgiliu. "Viewpoint: Space and Religion in Russia; Cosmonaut Worship to Orthodox Revival." *Astropolitics* 7, no. 2 (July 6, 2009): 150–63. https://doi.org/10.1080/14777620903113857.

Porter, Tenelle, Abdo Elnakouri, Ethan A. Meyers, Takuya Shibayama, Eranda Jayawickreme, and Igor Grossmann. "Predictors and Consequences of Intellectual Humility." *Nature Reviews Psychology* 1, no. 9 (June 27, 2022): 524–36. https://doi.org/10.1038/s44159-022-00081-9.

Powys, Llewelyn. *Llewelyn Powys: A Selection from His Writings*. MacDonald, 1952.

Rampley, Matthew. "The Ethnographic Sublime." *Res: Anthropology and Aesthetics* 47 (March 2005): 251–63. https://doi.org/10.1086/resv47n1ms20167669.

Reynolds, Gabriel Said. *The Qur'ān in Its Historical Context*. Routledge, 2008.

Rodríguez-Laiz, Antonio. "The Wright Brothers Weren't the First—or Were They?" *AERTEC*, December 8, 2023. https://aertecsolutions.com/en/2023/12/11/the-wright-brothers-werent-the-first-or-were-they.

Roth, Sol. "Towards a Definition of Humility." *Tradition: A Journal of Orthodox Jewish Thought* 13 (1973): 5–22.

Rowbotham, Samuel B. *Zetetic Astronomy: Earth Not a Globe*. Left of Brain Books, 2024.

Russo, Kate, and Andrew W. Bailey. "Deeper Insights into the Solar Eclipse Experience: Piloting Project Awe." *Celebrating the Wonder of Science in the Shadow* 56, no. 3 (March 8, 2024). https://doi.org/10.3847/25c2cfeb.530f7677.

Sagan, Carl. "Wonder and Skepticism." *Skeptical Inquirer* 19, no. 1 (1995): 24–30.

Sarma, Kamal. *Mental Resilience: The Power of Clarity; How to Develop the Focus of a Warrior and the Peace of a Monk*. New World Library, 2008.

Schlenoff, Daniel C. "The Equivocal Success of the Wright Brothers." *Scientific American* 289, no. 6 (December 2003): 97.

Shapshay, Sandra. "A Two-Tiered Theory of the Sublime." *The British Journal of Aesthetics* 61, no. 2 (February 8, 2021): 123–43. https://doi.org/10.1093/aesthj/ayaa047.

Shaw, Philip A. *The Sublime*. Routledge, 2017.

Shore, Paul. “Recent Studies in Jesuit History.” *Journal of Religious History* 31, no. 3 (August 16, 2007): 316–23. https://doi.org/10.1111/j.1467-9809.2007.00587.x.

Shweder, Richard A. “The Illusions of ‘Magical Thinking’: Whose Chimera, Ours or Theirs?” *HAU: Journal of Ethnographic Theory* 12, no. 1 (March 1, 2022): 319–25. https://doi.org/10.1086/719289.

Singer, Bayla. *Like Sex with Gods: An Unorthodox History of Flying*. Vol. 3. Texas A&M University Press, 2003.

Smith, Brian K. “Eaters, Food, and Social Hierarchy in Ancient India: A Dietary Guide to a Revolution of Values.” *Journal of the American Academy of Religion* 58, no. 2 (1990): 177–206. https://doi.org/10.1093/jaarel/lviii.2.177.

Smith, Jo Durden. *The Essence of Buddhism*. Arcturus, 2005.

Stephanson, Anders, and Eric Foner. *Manifest Destiny: American Expansionism and the Empire of Right*. Hill and Wang, 2005.

Stern, David P. “Robert Goddard and His Rockets.” *Robert Goddard and His Rockets*, 2005. https://pwg.gsfc.nasa.gov/stargaze/Sgoddard.htm.

St. Laurent, Simon. “Light and Silence: Reflections on Quakerism.” *Light and Silence* Blog. Accessed June 24, 2024. https://lightandsilence.org/2007/02/humility.html.

Suckow, Elizabeth. “NACA Overview.” NASA, 2009. https://www.nasa.gov/history/naca/overview.html.

Suebsaeng, Asawin, Tim Dickinson, and Andrew Perez. “Trump and Elon’s ‘Pointless Bloodbath’ at the FAA Is Even Worse than You Think.” *Rolling Stone*, February 21, 2025. https://www.rollingstone.com/politics/politics-features/trump-elon-musk-faa-air-planes-pointless-bloodbath-1235274324/.

Suenens, Kristien. *Humble Women, Powerful Nuns: A Female Struggle for Autonomy in a Men’s Church*. Leuven University Press, 2020.

Suzuki, Shunryū, Trudy Dixon, Huston Smith, Richard Baker, and David Chadwick. *Zen Mind, Beginner’s Mind*. Shambhala, 2020.

Swanson, Glen E. “The New Frontier: Religion in America’s National Space Rhetoric of the Cold War Era.” *Religions* 11, no. 11 (November 9, 2020): 592. https://doi.org/10.3390/rel11110592.

Tätte, Kai, Ene Metspalu, Helen Post, Leire Palencia-Madrid, Javier Rodríguez Luis, Maere Reidla, and Anneliis Rea. “The Ami and Yami

Aborigines of Taiwan and Their Genetic Relationship to East Asian and Pacific Populations." *European Journal of Human Genetics* 29, no. 7 (March 23, 2021): 1092–1102. https://doi.org/10.1038/s41431-021-00837-6.

Tavares, Frank, Denise Buckner, Dana Burton, Jordan McKaig, Parvathy Prem, Eleni Ravanis, and Natalie Trevino. "Ethical Exploration and the Role of Planetary Protection in Disrupting Colonial Practices." *Bulletin of the AAS* 53, no. 4 (March 18, 2021). https://doi.org/10.3847/25c2cfeb.cdc2f798.

Tetlock, Philip E. *Expert Political Judgment: How Good Is It? How Can We Know?* Princeton University Press, 2006.

Thaden, Louise McPhetridge. *High, Wide and Frightened*. Pathfinder, 2023.

Theilmann, John M. "Medieval Pilgrims and the Origins of Tourism." *The Journal of Popular Culture* 20, no. 4 (March 1987): 93–102. https://doi.org/10.1111/j.0022-3840.1987.9841353.x.

Thompson, Nainoa. "Reflections on Mau Piailug." *Proa Sailing*, July 29, 2010. https://starrigging.blogspot.com/2010/07/reflections-on-mau-piailug-by-nainoa.html.

Tiberghien, Guillaume, Raushan Mukhamedjanova, and Philip Feifan Xie. "Authenticity and Spectrality of Space Heritage: Baikonur Cosmodrome, Kazakhstan." *Tourism Geographies* 25, no. 5 (July 4, 2023): 1445–64. https://doi.org/10.1080/14616688.2023.2231422.

Traphagan, John W. "Religion, Science, and Space Exploration from a Non-Western Perspective." *Religions* 11, no. 8 (August 3, 2020): 397. https://doi.org/10.3390/rel11080397.

Trivers, Robert. "Reciprocal Altruism: 30 Years Later." In *Cooperation in Primates and Humans: Mechanisms and Evolution*. Springer, 2006.

Valdesolo, Piercarlo, and Jesse Graham. "Awe, Uncertainty, and Agency Detection." *Psychological Science* 25, no. 1 (November 18, 2013): 170–78. https://doi.org/10.1177/0956797613501884.

Van Riper, A. Bowdoin. *Rockets and Missiles: The Life Story of a Technology*. The Johns Hopkins University Press, 2007.

Van Tongeren, Daryl. *Humble: Free Yourself from the Traps of a Narcissistic World*. The Experiment, 2022.

Watterson, Bill. *The Days Are Just Packed: A Calvin and Hobbes Collection*. Andrews and McMeel, 1993.

Weibel, Deana L. *A Sacred Vertigo: Pilgrimage and Tourism in Rocamadour, France*. Lexington Books, 2022.

Weibel, Deana L. "Black Ugliness and the Covering of Blue: William Shatner's Suborbital Flight to 'Death.'" *The Space Review*, October 18, 2021. www.thespacereview.com/article/4265/1.

———. "The Highs and Lows of Extreme Tourism: The Titan Accident and Commercial Expeditions to Space and the Deep Sea." *The Space Review*, July 31, 2023. www.thespacereview.com/article/4630/1%E2%80%8B.

———. "The Sacred Moon: Navigating Diverse Cultural Beliefs in Lunar Missions." *The Space Review*, January 29, 2024. www.thespacereview.com/article/4733/1.

———. "Sherpas on the Moon: The Case for Including 'Native Guides' in Space Exploration." In *Reclaiming Space: Progressive and Multicultural Visions of Space Exploration*. Oxford University Press, 2023.

———. "Space Exploration as Religious Experience." *The Space Review*, August 21, 2017. https://www.thespacereview.com/archive/3310-1.html.

Weibel, Deana L., and Glen E. Swanson. "Malinowski in Orbit: 'Magical Thinking' in Human Spaceflight." *Quest: The History of Spaceflight Quarterly* 13 (2006): 53–61.

Weinberg, Albert K. *Manifest Destiny: A Study of Nationalist Expansionism in American History*. The John Hopkins Press, 1935.

Weitekamp, Margaret A. *Right Stuff, Wrong Sex: America's First Women in Space Program*. The Johns Hopkins University Press, 2006.

Wells-Jensen, Sherri, J. S. Johnson-Schwartz, and A. J. Link. "3.A. Para-Astronauts." eGrove. Accessed May 26, 2024. https://egrove.olemiss.edu/ssocia/2022/schedule/11.

White, Frank. *The Overview Effect: Space Exploration and Human Evolution*. Multiverse Publishing, 2021.

Whitman, Walt. *Leaves of Grass*. Rome Brothers, 1855.

Williams, Mark. *Goldilocks Planet: The 4 Billion Year Story of Earth's Climate*. Oxford University Press, 2013.

Wilser, Jeff. *The Explorers Club: A Visual Journey through the Past, Present, and Future of Exploration*. Ten Speed Press, 2023.

Wilson, David Sloan, and Edward O. Wilson. "Rethinking the Theoretical Foundation of Sociobiology." *The Quarterly Review of Biology* 82, no. 4 (December 2007): 327–48. https://doi.org/10.1086/522809.

Wired. "How Mapping Indigenous Knowledge Is Helping Nomadic Communities to Fight Climate Change—and Extinction." *Wired*, June 15, 2022. https://www.wired.com/sponsored/story/3d-mapping-indigenous-climate-change-extinction.

Wood, Danielle, Prathima Muniyappa, and David Colby Reed. "Opportunities to Pursue Liberatory, Anticolonial, and Antiracist Designs for Human Societies beyond Earth." In *Reclaiming Space: Progressive and Multicultural Visions of Space Exploration*. Oxford University Press, 2023.

Worth, Eve, and Aaron Reeves. "'I Am Almost the Middle-Class White Man, Aren't I?' Elite Women, Education and Occupational Trajectories in Late Twentieth-Century Britain." *Contemporary British History* 38, no. 1 (October 13, 2023): 71–94. https://doi.org/10.1080/13619462.2023.2265815.

Wrangham, Richard W. *Catching Fire: How Cooking Made Us Human*. Basic Books, 2010.

Yang, Bo, Yeyao Liu, and Jiawei Liao. "Manned Submersibles—Deep-Sea Scientific Research and Exploitation of Marine Resources." *Bulletin of Chinese Academy of Sciences* 36, no. 5 (May 20, 2021): 1–8.

Yong, Ed. *I Contain Multitudes: The Microbes within Us and a Grander View of Life*. Ecco, 2018.

YouTube. "Earthrise: A Conversation with Apollo 8 Astronaut Bill Anders (Official NASA Video)." April 21, 2023. https://youtu.be/uFfFsOu7yqY?si=oluZ5DD6Whxk8wcF.

———. "NASA: Earthrise: The 45th Anniversary." December 20, 2013. https://youtu.be/dE-vOscpiNc?si=EjEvifWZQiBz_bc7.

Index

Founded in 1893,
UNIVERSITY OF CALIFORNIA PRESS
publishes bold, progressive books and journals on topics in the arts, humanities, social sciences, and natural sciences—with a focus on social justice issues—that inspire thought and action among readers worldwide.

The UC PRESS FOUNDATION
raises funds to uphold the press's vital role as an independent, nonprofit publisher, and receives philanthropic support from a wide range of individuals and institutions—and from committed readers like you. To learn more, visit ucpress.edu/supportus.